RETRAIN YOUR ANXIOUS BRAIN

Also by
JOHN TSILIMPARIS

Mortal Bonds: A Memoir

RETRAIN YOUR ANXIOUS BRAIN

Practical and Effective Tools to Conquer Anxiety

JOHN TSILIMPARIS, MFT
AND DAYLLE DEANNA SCHWARTZ

HANOVER
SQUARE
PRESS

HANOVER SQUARE PRESS™

Recycling programs for this product may not exist in your area.

ISBN-13: 978-1-335-42569-0

RETRAIN YOUR ANXIOUS BRAIN

First published in 2014. This edition published in 2022 with revised text.

This edition published by arrangement with Harlequin Enterprises ULC.

Library of Congress Cataloging-in-Publication Data has been applied for.

Hanover Square Press
22 Adelaide St. West, 41st Floor
Toronto, Ontario M5H 4E3, Canada
HanoverSqPress.com
BookClubbish.com

Printed in U.S.A.

For my mother and father.

"I am bigger than anything that can happen to me.
All these things, sorrow, misfortune, and suffering, are outside my door.
I am in the house, and I have the key."

—CHARLES FLETCHER LUMMIS

"Our anxiety does not come from thinking about the future,
but from wanting to control it."

—KAHLIL GIBRAN

"A mind is like a parachute. It doesn't work if it is not open."

—FRANK ZAPPA

CONTENTS

INTRODUCTION

Everyone experiences symptoms of anxiety throughout their lives at different times and in varying degrees. Some people shrug off or dismiss everyday stress and anxiety as part of life's unavoidable ups and downs and don't think much of it beyond that. But for others, ongoing stress leads to developing symptoms of chronic anxiety that can significantly impair people's daily functioning and cause a major disruption in their lives. Anxiety often sneaks up on people and feels unavoidable. Fortunately, anxiety can be treated, but the consequences for allowing it to go unchecked can have negative results.

It's estimated that about 40 million adults in the United States suffer from some form of anxiety. Yet despite how prevalent it is, a large percentage of anxiety sufferers don't seek help because they feel too uncomfortable or embarrassed to talk about it, or they're scared that admitting to having it will stigmatize them. To many people, especially those who haven't experienced it over a long period of time, anxiety seems like a self-inflicted ailment that people need to get over. Because it has often been negatively viewed as a sign of weakness, having anxiety also causes many individuals a great deal of personal shame, which can lead to suffering alone. It can be a painful way to live. I know, because it happened to me.

MY EXPERIENCE WITH ANXIETY

I didn't write this book about handling anxiety just from the point of view of a therapist who learned about it in school, although I am one. This is *my* story, too. I know about anxiety on a personal level because

I have a long history with it dating back to childhood. I wrote this book because after many years of suffering, I learned practical, effective ways to handle my anxiety that allowed me to finally enjoy my life, and I want to pass them on to help you learn how to retrain your anxious mind. I write from both the perspective of a therapist and as someone who lived through many years of debilitating anxiety, and it was by using the tools in this book that I was finally able to manage it.

I know what it's like to wonder if you're going crazy because of the anxious feelings that come over you for no discernible reason and that no one understands. I can relate to the powerless feeling you can have when you can't control your escalating panic and don't know how to find relief. But I also know the relief of finally learning how to help myself. You can do it, too! I'll help you, just as I did when I treated people with anxiety on the TV show, *Obsessed*. I heard from people around the world who saw me on the show, asking if I could share more of my techniques. They also suffered from anxiety and were desperate for help to control it. I relate to their pain, and if what I share makes a positive difference for you, it matters to me.

My anxiety began when I was young. By the time I was eight years old, I was aware that I was a different kind of child. I just wanted to fit in like everyone else, but I knew that I didn't. No matter how hard I tried, I felt alienated by my puzzling condition that no one could understand, not even me. I reacted to situations in ways that seemed wrong or unacceptable and didn't know why. I was criticized or ignored when I reached out for support. For years I wondered what would become of me and how I'd ever be able to survive in the world with such an incapacitating handicap. The uncertainty carved deep grooves of permanent insecurity. The future looked grim and scary.

It all started one night when I suffered my first anxiety attack after my parents put me to bed. Out of nowhere, a terrifying current of fear surged through my body like a demonic possession. My heart began beating rapidly, I gasped for air and I began to sweat. My mind raced uncontrollably, and I thought I was going to die. I ran into my parents' bedroom and, in a frantic state, I woke up my mother, not knowing what was happening to me. She lifted her head from the pillow and in a drowsy tone whispered very softly that I should be careful not to wake up my father. It wasn't until I was in my mid-twenties that I realized how significant her words would become.

She murmured, "Go back to bed. You have nothing to be afraid of." I went back to my bed as she instructed and hoped she was right. For a moment I believed her without question, but the panic continued. As I lay awake for the rest of the night, trembling, I wondered if I was going to see the daylight again. I felt alone and terrified, knowing I had no one to turn to. As a child, it's especially hard to understand what's going on when anxiety hits, which makes you feel powerless. I wasn't in control of myself—the anxiety was. It was the first of many dark and lonely nights to come.

For the next four years, the panic attacks and general anxiety that developed from them occurred intermittently and without warning. I gradually understood that something was amiss—because if there was nothing to be afraid of, as my mother declared, yet I was still afraid, then the only answer could be that something was truly wrong with me. I also remember thinking that I must be stupid and weak if I was scared of nothing. There was no identifiable source of stress that I could verbalize to my mother—no monsters under the bed or a boogeyman in the closet. There was simply no reason to feel so frightened. I had never even heard of the word *anxiety*. No one ever spoke about it to me. I don't think anyone in my family even thought about it. My problem was just shrugged off, and I was left to deal with it on my own.

By the age of twelve, my panic attacks mysteriously went dormant, and for the next nine years, they stayed hidden in the deep recesses of my mind. But when I returned home to New York City after college, the anxiety returned with great intensity. By then I was an adult and had a more mature ability to understand that I needed help. I also seriously recognized my need to seek help instead of letting my problem be minimized or ignored by my parents. Desperation finally convinced me to give in and go to a psychotherapist. At first my parents tried to deter me from seeking any kind of psychiatric care. They believed that therapy was only intended for crazy people and shamelessly admitted they were worried that the therapist would somehow brainwash me and turn me against them.

Usually, their opinions—especially my father's—were extremely influential and difficult to ignore without fear of reprisal. Luckily, I didn't listen to them this time and moved forward with treatment despite their objections. I was desperate and knew I needed someone else to guide me out of my confusion. It was one of the most crucial decisions I ever made in my life.

Many people who suffer from anxiety are often discouraged from getting help by well-meaning but uninformed people. The belief that anxiety can be controlled if you simply try hard enough often prevents people from taking you seriously and can make you feel low about your inability to "get over it." It also can deter you from taking therapeutic steps to actually get the kind of help you need.

After two years of on-and-off treatment sessions with a very bright and compassionate therapist, I learned to use solid coping tools to help manage my anxiety better. However, I also realized that despite the therapy I still felt very different from other people. I noticed I was still more reactive to potentially stressful situations and worried a lot about things that most people easily accepted.

I began to understand that it was critical to develop ongoing maintenance of my symptoms in order for me to function normally since they kept coming back. Unlike the average person, life's ordinary twists and turns were sometimes hard for me to navigate because of my susceptibility to stress. The good news was that every time the symptoms returned, they were markedly less intense than before and lasted for a shorter duration. The panic attacks occurred less frequently, and the lingering anxiety was manageable. I happily recognized that my progress was astounding. The future didn't look so grim anymore.

A big part of an anxiety sufferer's angst can be a lack of understanding. Have you ever been asked, "Why are you driving yourself crazy?" when someone observes you in an anxious state and assumes you're to blame? That can make you feel like you're going crazy, because people don't take seriously the scary things you're experiencing and you don't know where to turn to get answers. When you have no idea about what's going on in your head, it creates a helpless feeling. That's why after therapy I felt such a relief to have something to call it—anxiety— a diagnosis I could reckon with that was treatable.

I finally began to accept anxiety as part of my life. Ultimately, I let go of feeling stigmatized for having this condition and stopped seeing myself as a weak person; instead I began referring to myself as someone who suffered from an anxiety disorder. It was very liberating. But the most pivotal thing I learned about my anxiety was how my thoughts severely skewed my perception of situations and altered the way I was feeling. I realized that for most of my life I magnified things that didn't go well to a catastrophic degree and thought in all-or-nothing terms. I discovered I was a perfectionist about anything related to performance, image and how I presented myself to others. This kind of insight began to change my life in immeasurable ways.

I also recognized that growing up with a very critical father made me afraid to disappoint people and I wanted everyone to like me. And most important, I often fell victim to the illusory trap of thinking I could control everything important to me. So when I did try to control things and my efforts failed—which was most of the time—it made me feel worse, like I was trying to catch the wind with a butterfly net. Furthermore, I finally understood that telling a person who suffered from chronic anxiety that they had "nothing to be afraid of" the way my mother did years earlier was like telling a severe alcoholic to "just stop drinking." It was not that simple. But my mother, who was actually a very loving woman, was incapable of fully comprehending the depth of my condition.

Over time, I learned to listen only to the intent in her words and not to the details of them. She tried to help in the best way she knew how, and knowing that further helped me gain a healthier perspective about my earliest years suffering with anxiety. Throughout my late twenties and thirties I worked on my coping skills very diligently and managed to keep my symptoms at bay for the most part. Every now and then, a life stressor spiked my anxiety for a brief spell, but there were no major incidents like in the past. Then, after my divorce in 2000 and the deaths of both my parents in 2004, I was challenged by extreme bouts of anxiety once again. But I recovered quickly thanks to the effective skills I had mastered.

TAKING ANXIETY SERIOUSLY

The shame I once felt about my anxious responses to life and my ignorance about why they were happening had kept me from getting help for many years. This has been a consistent problem for many people who suffer from anxiety. Fortunately, anxiety's negative reputation is changing. Many more people are seeking help and coming to treatment for anxiety than ever before, likely due to the fact that, today, mental health issues are slightly less stigmatized than in the past and there's a better neurobiological understanding of how our brains work when we are affected by anxiety.

But make no mistake—for many, anxiety as well as other mental health–related issues are still not respected by mainstream society as legitimate conditions to be reckoned with. Anxiety in particular is still considered by many to be the result of human weakness. Furthermore, with all the recent turmoil and uncertainty of the last few years, including persistent fear and stress due to the global pandemic and a host of other real and foreseeable worries, the number of people suffering from anxiety will rise exponentially. You can't listen to those people who tell you to stop driving yourself crazy and to start coping better.

Anxiety is a real clinical disorder, whether the general public accepts it or not. And there are effective ways you can control it that I'll share with you. When you acknowledge that you have a problem with anxiety, you can use the tools in this book to help you learn how to make peace with it and adopt a different way of handling the triggers that bring it on. While you may always have a tendency to be anxious, you can develop coping mechanisms that can keep the severe reactions at bay like I finally did.

Retrain Your Anxious Brain is written as a survival guide for anyone who currently suffers from anxiety—from the mild, everyday anxiety to the severe and debilitating type. This book is not meant to be a cure for anxiety, nor is it meant to be a quick-fix antidote to life's problems. Instead, it's a design of balanced thinking that focuses on improving the inner management of yourself. The book offers "sustainable" forms of anxiety reduction skills because once you learn the techniques prescribed, you can replenish them whenever you need to. In a sense, you can be your own natural resource. Most books on anxiety have tools to help you cope only with the *effects* of the disorder. I provide a variety of ways to identify and short circuit what *causes* it so you can curb or lessen it from the get-go and have fewer negative effects.

The theory behind my approach to treating anxiety is that you can change your internal responses to the things that make you anxious and scared. By doing so, you'll start to feel better as your new ways of reacting to situations that trigger anxiety become more comfortable and automatic. This can be much more effective than trying to change what's going on in the world itself and staying anxious. I'll show you life-adaptability tools that can help you adjust and regulate yourself when the difficult and unavoidable aspects of being human occur. For example, if spending a day with your family triggers stress about not having accomplished as much as your siblings and you feel inadequate, you'll learn how to handle those feelings in ways that don't escalate anxiety.

This design works by rejecting the belief that you need to focus on and change the external circumstances of your life in order to reduce anxiety, especially since so much of what you encounter is beyond your control. Instead, I encourage an internal focus that holds you accountable for making changes. This method promotes a major inner shift in how you react and subsequently respond to anxiety-provoking situations. The book will teach you how to adapt to any aspect of life that involves a stressful change—such as experiencing a loss, illness or occupational stress—by simply reevaluating and eventually altering the fixed thoughts that you have inside you that you've previously used to deal with anxiety. I'll show you how to effectively create and establish your own personal reality of life that allows you to feel more in control of your anxiety.

Retrain Your Anxious Brain isn't for people who are looking for swift results or solutions to problems. The main goal is to help you learn how to streamline your thinking about the problems you encounter. I will guide you in finding ways to reevaluate who you are in relation to the world, how you relate to other people, and how you deal, or don't deal, with the difficult times in your life. If you shift your perspective and interpretation about how the things that give you anxiety are outside of your control, you'll gradually be able to change your response to them.

Since I'm no stranger to suffering from anxiety, the viewpoint of anxiety management that I present includes some of what I learned from getting my own treatment, studying what works for treating anxiety and working with many patients. The techniques I share have proven to be very effective. They're partly ingrained in existing theories such as cognitive behavioral therapy (CBT), rational emotive therapy (RET) and existential therapy.

CBT and RET are skills-based therapeutic styles that are instructive in nature but also highly user-friendly because they center on the anxiety sufferers' negative thinking patterns instead of their character. They also focus on helping to identify and change distorted thought patterns. The ultimate goal of CBT and RET is to help restructure your thinking so you can learn how to separate realistic thoughts from unrealistic ones, like worrying that because your boss looked upset when you arrived to work this morning means he's unhappy with your job performance and you may be fired. These therapies will also help you understand that your feelings are based on your own personal beliefs and interpretations, which may not be the true reality of what you experience. Both CBT and RET posit that thoughts, feelings and actions have a reciprocal relationship with one another. Therefore, if you restructure your inner thoughts and train yourself to identify negative ones that will trigger anxiety, it will inevitably lead to a positive change in your behavior.

As per the existential theory, you are the architect of your life, and who you are and what you have become is a product of your choices and actions. You alone have the responsibility of your life, and you're not always controlled or determined by external forces. Your existence is never fixed, and you're forever recreating yourself by living in a constant state of transition—an ever-shifting attitude of questioning, learning and evolving. This is why existentialists also believe that you alone make your own subjective meaning of the world. My coping techniques will allow you to have more control over how you respond to your life's ups and downs.

Throughout the book you'll have the opportunity to create a personalized plan for handling your anxiety by recognizing the triggers that make you anxious. I'll give you easy and practical exercises to do and questions to ask yourself that will help you begin the process of retraining your anxious mind and positively reflecting on your life, using my anxiety-reduction model. The tools and exercises will give you the kind of insight that will allow you to recognize and define your triggers, categorize them and begin to shift into alternative ways of thinking about what's happening around you. Then you can begin to create your own reality—a chief necessity for calming an anxious mind.

MOVING FORWARD

Today, as a psychotherapist specializing in anxiety disorders, I consider myself fortunate to have used my own personal history with anxiety as inspiration to help others in need. If anyone would have told me years ago that one day I'd be helping people with the same condition I struggled with, I'd have laughed in their faces. I never considered the possibility that I could turn my anxiety into a positive.

It seemed inconceivable. But, ironically, my suffering became my teacher and my healing became my mission. It all worked out as my misery was validated and turned into a blueprint for others to create their own plan to retrain their anxious minds.

This book is a kind of reflection of the very same process of healing I experienced over my last thirty years of being both a patient and a mental health professional. It presents many of the coping skills and anxiety-reduction techniques I used that were helpful for my own recovery, including the many psychological insights that changed my life forever.

The distinctive therapeutic concepts offered in this book were envisioned and later adapted and implemented successfully over many years of treating patients with all types of anxiety, and for hundreds of people, both personal patients and people I treated for anxiety during my appearances on *Obsessed*. While they are simple, basic, common sense insights, they also have the power to alter your life by offering a new way of thinking and relating to the world, and these tools are available to anyone who is willing to try them.

I hope that after you read this book you won't feel alone because of your anxiety or that you're different from most people, as I did when I first realized I had a problem. The truth is, you're not alone. And although you may be different in many positive ways, you're *not* different because you suffer from anxiety—at least not anymore. There is indeed life during and beyond anxiety. I am living proof. You can retrain your anxious mind. If I can do it, you can, too!

Chapter 1

UNDERSTANDING
ANXIETY

"Anxiety is the hand maiden of creativity."

—T.S. Eliot

Anxiety is one of the most misunderstood conditions of the human mind. Yet anxiety disorders are the most common of mental illnesses in the United States. It's estimated that about 18% of the population suffers from some form of anxiety. Anxiety disorders are real, not something that fragile people bring on themselves. In this chapter I will help you to understand the myths and truths about anxiety so that you can do more to take charge of yours, or so you can help someone you care about who suffers from it.

WHY ANXIETY IS OFTEN KEPT QUIET

Even though anxiety is so prevalent, studies show that many people who suffer from it often remain in what I refer to as the "angst" closet for years before reaching out for professional help. Many anxiety sufferers feel different levels of shame or they are scared people will ridicule them if their secret is discovered. I used to feel that way when my mother refused to acknowledge my problem as real. After all, how could I share it with others if my own mother acted like I had no problem and it was all in my head? It's a common response from people who don't understand anxiety. We've come a long way in terms of lowering the stigma of having anxiety and other mental health conditions as a whole in more recent years, but a lot of work still needs to be done.

It's still painfully apparent, and at times shocking, that in this day and age, anxiety, which is often an unbearable condition involving the mind, continues to be viewed by many as an affliction that only affects inferior people. Perhaps there is sanity in numbers. For example, when was the last time you heard someone being judged or ridiculed for having some kind of physiological condition? Apparently, over time, there is a rational and collective acceptance about any human condition that has evidence showing that multitudes of ordinary people are stricken with it every year. But it's very different for illnesses of the mind, including anxiety.

How can this be changed? The more anxiety is discussed as a legitimate condition that people don't bring on themselves, the more people will have empathy, not negative reactions, to someone struggling with it. If people from all walks of life—from college students to seasoned professionals, from blue-collar workers to white-collar executives— admit to suffering from anxiety and getting help for it, the more the average person will accept it like they do other conditions.

My hope is that this chapter and the entire book itself will help shed a more positive light on anxiety and help to validate the seriousness of this condition. If having anxiety becomes accepted like other conditions are, future generations will finally receive the long-awaited recognition that anxiety affects many average people and that it's merely a genetic and environmentally influenced variant of being human. And, there's help available to allow you to manage it.

THE BENEFITS OF ANXIETY

To understand the many aspects of anxiety, you must first recognize some of its benefits. Anxiety is an adaptive function of the human body that exists for the purpose of protecting us from harm.

Anxiety is part of an evolutionary response mechanism developed slowly over many millennia. If primitive humans had no internal alarm system when the dangers of ferocious animals lurked outside their cave looking for their next meal, they wouldn't have survived.

Their life-preserving arousal mechanism, better known as the fight-or-flight response system (also called the fight-or-flight-or-freeze response system), is a biochemical reaction that was developed to help the body sense and respond quickly to danger. We could have become extinct had this system not been part of our makeup. But thanks to our hairy, cave-dwelling ancestors, generations have inherited this fundamental impulse. It's not needed today the way it used to be, but it's wired into us. Some people's systems are more sensitive than those of others.

In essence, anxiety is a response or a reaction to something that the brain perceives as dangerous or threatening. Without anxiety, you might walk through life and go about your business naively, in a blissful state, and be oblivious to simple dangers. A car may strike you because you're not paying attention while crossing the street. Or you may become a negligent parent by not safeguarding your home properly if you have young children. Or you may underperform at work because you are not worried about losing your job. A tendency to feel anxious in specific situations does help keep you on your toes when it comes to protecting yourself and your loved ones.

Many artistic people, such as writers, actors, dancers and other artists, have credited the feeling of being anxious about their work and the pressure to perform well as triggers that spark their creative fires. Even great intellectuals and inventors throughout history have professed to having been indirectly inspired by the fear of failure or, most commonly, the fear of humiliation or ruin. Being anxious at times means you care about your life and the meaningful people in your life. It means you're a responsible person who wants to succeed and provide for yourself and your family.

Without anxiety, reaching goals and conquering risks wouldn't be so valuable to you. In a sense, anxiety gives you purpose and drive.

It keeps you on the right path and helps you appreciate the good you have. But for the 18% of us who suffer from excessive amounts of it, this organic boost of angst can be very incapacitating. Too much anxiety can cause so much distress that you can't function properly. That can lead you to fail at fulfilling major role obligations in your life that most people take for granted, such as taking care of your children properly, going to work every day and attending school. Anxiety can be so distressing that sufferers would give anything to be rid of all anxious feelings for good. But when you can understand and appreciate its positive effects better over time and learn how to harness its returns, it can seem less like something that only works against you.

UNDERSTANDING THE FIGHT-OR-FLIGHT-OR-FREEZE RESPONSE

The fight-or-flight response is an automatic inner alarm system that physically prepares the body to attack and defend itself (fight), or to run away and protect itself (flight). The freeze aspect describes the paralyzing effect that some people experience when they're so scared that they feel frozen in terror. It also refers to the body's instinct of staying still or "playing dead" in order to ward off a predator. When you experience something that feels dangerous or threatening, a warning bell is triggered. Neurotransmitters are released in the brain and send messages to the adrenal glands, which produce powerful hormones such as cortisol, which is called the "stress hormone" because it's found abundantly in the bloodstream of anxious individuals. What follows as a result of these hormones surging through the body are highly disquieting physical responses that don't feel good. They can also be very scary, which adds to the anxiety that's already working overtime.

The typical symptoms that you might experience from this flooding of hormones can be very troubling if you don't understand them. You may suddenly experience a rapid heart rate, shallow and labored breathing, sweating, and your mind can start to race uncontrollably, scanning the horizon for seen or other unforeseen dangers.

The rapid heart rate caused by the fight-or-flight response produces an increase in the strength of the heartbeat for good reason. It's critical in your body's preparation for fight-or-flight. Its purpose is to pump blood quickly to the areas of the body that might be needed to face a fight or run away fast, including the large muscle groups, especially your arm and thigh muscles. Blood is then diverted from peripheral areas like your fingers and toes because if the body is badly injured, it's less likely to bleed to death. This is why people who suffer from anxiety often report having clammy hands and tingling sensations in the feet and toes.

The shallow, labored breathing that's often experienced also has a pertinent function. The pronounced and enhanced breathing helps to prepare you for the fight or to take flight by bringing more oxygen to your body. While this response is supposed to be for your own good, some of the side effects can get intense. The shallow breathing can escalate to hyperventilation, which can scare you. This can also give you choking sensations and chest pain. And because the blood to the head is decreased by excessive panting and gasping for air, you can also feel dizzy or light-headed.

The fight-or-flight response mechanism produces an increase in perspiration to keep the body from overheating while it's in action. And an increase in sweat helps make the skin slippery and more difficult to latch onto if a predator catches you.

But the chief function of the fight-or-flight response is to reliably alert the mind to a threat or danger in the area. The mind immediately shifts its focus and redirects its attention to the immediate surroundings to check for danger. Because of this, some individuals have difficulty with their memory and trouble concentrating and keeping their attention on the present. Or the mind can shift into freeze mode where it goes completely blank, leaving you feeling helpless about what course of action to take. All of this is wired into your system to protect you as it did our cave-dwelling ancestors. While the extent of the fight-or-flight response isn't necessary today, it's there whether you need it or not, and it is the trigger for your anxiety.

WHO IS AFFECTED BY ANXIETY?

One of the many problems with suffering from this condition is that the anxious mind struggles to differentiate between a warranted panic situation, like being chased by a hungry grizzly bear in the woods, and the typically mild stress of something basic, like being late for an appointment. The anxious, primitive brain does not distinguish between stress triggers that easily, and every stressor can become a catastrophe in the making.

Most people who suffer only mildly from worry and the daily stresses of life—people who make it through a regular day relatively unscathed emotionally—understand anxiety in a way much like how they experience fear when they're watching a horror film. When watching one, they know it's just a movie and that they're watching fiction. Yet they still get scared as if it were really happening when they see it on the screen. They jump out of their seats or cover their eyes. Once the movie is over, they restabilize to a calmer disposition, and their fear-arousal symptoms subside. They return to their normal activities of living without the anxiety they felt while watching the film.

But if you worry excessively, you don't experience anxiety as a brief and occasional phase. It takes more of a hold on you, making it harder to restabilize. One seemingly innocuous event can trigger your anxiety and become debilitating. Despite having suffered for many years with the same recurring fears, the brain doesn't always learn that there's nothing terrible to be afraid of. It's been found that the fight-or-flight response system that's wired to protect you actually sidesteps the critical part of the brain where you store thoughts—the area that controls how you interpret and rationalize stress triggers.

Another reason for why some people are more at risk than others for being crippled by anxiety is because of their environment. A traumatic childhood, accident, injury, chronic illness, death or other incidents that may have scarred you have the potential to leave indelible psychological consequences, which influence your reactions in the future. A person who grows up in an unstable, chaotic family where he or she is exposed to physical or verbal abuse is likely to become more susceptible to anxiety symptoms—or even more likely to have an anxiety disorder—than someone who came from a stable home. And past experiences can manifest in current situations that are similar. For example, people who had a humiliating public speaking experience are likely to get more nervous in the future when they're communicating in a group or social setting than other people do.

Genetics also plays a role in the neurobiological makeup of your brain. Studies show that the psychological susceptibility discussed earlier can be transmitted multi-generationally or passed down to offspring. Mom's and Dad's experiences in life can begin a genetic legacy of angst—that angst could be passed down from grandparents or their parents. It can be difficult to try to pinpoint exactly where your anxiety came from, but trying to identify the cause can give you more understanding and enable you to take control of it.

My father was, by definition, an anxious and worrisome man who never sought treatment or help of any kind. Instead, he displaced his fears onto his wife and children. He was often punitive and unfeeling in the way he disciplined my brother and me, and he treated my mother very disrespectfully. Because he was also prone to fits of rage, his mood swings led to verbal and sometimes physically abusive behavior. As I mentioned in the Introduction, I've suffered from anxiety with a history of severe panic attacks throughout my life, beginning when I was about eight years old. While in therapy in my twenties, I discovered that I, quite possibly, not only inherited my father's anxious psychological makeup, but I was also a victim of his abuse. I didn't connect the dots until later on in life.

To this day I continue to unravel and piece together how and why I often react to things the way I do. It's either genetic, or it's the prolonged exposure to my father's behavior over many years that wired me for a life of oversensitivity, or it's both. My younger brother, who was exposed to the same trauma and instability in our home, has never experienced the types of psychological challenges I have. He was somehow spared the genetic legacy, which, as a result, may have rendered him immune to the potentially damaging trauma. Two boys growing up with the same abusive father and the same chaotic environment, but only one ended up being an anxious person. The dynamics of the brain can be very interesting and puzzling as well.

My brother and I have a great deal in common but when it comes to how we process and react to stimuli in our lives, we couldn't be more different. I tend to be a sponge, often absorbing the stresses of the day and feeling weighed down by them. My brother is the exact opposite. He's one of the most temperate people I've ever met despite having been raised in the same crazy household. In short, genetics and environment can be key factors that contribute to why someone is more likely to be affected by serious anxiety. But those factors don't necessarily wire everyone for a life of having to deal with it.

STRESS VS. ANXIETY

To fully understand the anxious mind, it's important to make a distinction between stress and anxiety because often the terms are used interchangeably. Stress and anxiety typically share similar characteristics in the physical symptoms they trigger, such as rapid heartbeat, labored breathing, muscle tension, restlessness, irritability, fatigue and sleep disturbance, as well as cognitive symptoms, such as difficulty concentrating, racing thoughts and excessive worry. However, anxiety is a mental disorder, and stress is not.

In the last twenty years of treating individuals with both stress and anxiety, I've recognized some clear differences between the two. Stress tends to be an edgy, worried reaction to an identifiable life issue, such as divorce, a relationship problem, illness in yourself or someone you love, unemployment, financial hardship or moving. The stressful feelings typically dissipate once the issue has decreased or been resolved. While the stress itself is still part of the fight-or-flight response mechanism that kicks in as a reaction to specific situations, it's considered a more moderate type of arousal compared to an anxiety or panic attack.

Stress also manifests itself as more of a persistent worry about specific things than an anxious impending doom. The duration of stress is typically shorter. Another important aspect of stress is that it doesn't usually cause as much impairment in the day-to-day functioning of a person's life like anxiety does. Furthermore, many people often experience what we call "healthy stress." This is the kind that motivates and sometimes inspires the pressure you may put on yourself to perform well at school, to excel at your job, to pursue a goal that matters to you, and to protect yourself and your loved ones. A little healthy stress is needed to survive in the world.

Anxiety is more of a response to life issues or other triggers that comes from being scared. It's also considered to be a reaction to something that stimulates stress and then sparks the fight-or-flight response mechanism, but these triggers are not always obvious, so it's hard to identify them. And, as discussed earlier, at times the source of anxiety may be perceived in a way that's out of proportion to the actual reality of the situation that scares you. Unlike stress, which has a specific cause and typically ends when there's a resolution to what causes it, anxiety can be a chronic and collective general feeling of discomfort that has built up over months or years. It's not limited to a finite amount of time that only lasts as long as the specific stress episode. Further, unlike stress, anxiety manifests more intensely than just a persistent worry. It can come on more like an ongoing feeling of panic and dread with no end in sight. When this happens, you probably won't be able to identify the trigger or reason that causes it.

Anxiety may cause people to experience serious cognitive disturbances that include creating thoughts that take a fearful, catastrophic or irrational direction. This can create distorted beliefs and a general doom-and-gloom type of outlook about what's happening in your life. Many people with anxiety also suffer from general disturbances to their daily routines. Anxiety is indeed responsible for causing significant impairment in how people function in social situations and at work on a day-to-day basis. As a consequence of those disturbances, an anxious person sometimes has to make painful sacrifices, which can begin a domino effect of shame and self-loathing.

For example, some individuals' symptoms are so acute they have to quit their jobs, give up traveling, or avoid driving on freeways or riding in elevators, or curtail doing many other things that most people consider to be normal activities. Even thinking about these stressors can send them into panic mode and trigger behavior that others won't be able to understand. When this happens, your self-esteem goes way down and you may avoid people who will want to know what's going on. It's hard to explain what's going on to someone else when you don't quite know why yourself, which adds another level of embarrassment and frustration. And it can be hard for people who don't understand the difference between suffering from anxiety and assuming they're just "stressed out" to look for real help.

For example, I once treated Emma, a newly enrolled graduate student who reported feeling stressed about flunking out of school on a daily basis. She experienced excessive worry throughout the day that prevented her from concentrating on her schoolwork. She also suffered from insomnia, poor appetite, difficulty breathing and racing thoughts about disappointing her family. Emma even reported that the "stress" was affecting her desire to socialize with friends and that she felt very inadequate about not dealing with the demands of graduate school. She often said, "What's wrong with me? I should be handling this better." When I asked her how long this had been going on, she revealed a long history of being worried and anxious about many things in her life, especially in relation to performing well in academics. She also revealed feeling like a loser whenever she was challenged by life events in this way.

While working with Emma as my patient, I helped her to understand the differences between stress and anxiety. I also helped her to realize that her symptoms were clear signs of moderate to severe anxiety, which needed attention immediately. If she continued to ignore or minimize them, they could have escalated and possibly caused her to be unable to function and get through the school year. Emma found it eye opening and a relief to know it wasn't her fault.

Once she grasped the understanding that she was suffering from a serious condition called *anxiety* instead of just stress, she felt less critical of herself and was then able to be more compassionate with herself about what she was going through. It allowed her to work on getting well instead of trying to dismiss her feelings. The diagnosis assisted her to reckon with her anxiety better and take it more seriously. After a few months of intensive therapy, Emma learned coping tools, which helped her to de-escalate her symptoms, slow down her racing thoughts and stabilize herself enough to get through the school year.

Like me, many of the individuals who are more prone to anxiety than stress meet the criteria to be diagnosed as having an anxiety disorder. Ironically, recognizing that you may have an anxiety disorder can give you a sense of relief because it often justifies the suffering you've had and validates the years of pain and confusion you've endured. The diagnosis provides a stamp of approval that the symptoms you've suffered are part of a clinical illness that requires attention. It confirms that what you've experienced for a long time is not simply the temporary rash of the nerves that many people try to assign to your condition. Getting a clinical diagnosis also tends to inspire hope that your anxiety can be treated and shows you that your symptoms are recognized as a condition that many people suffer from. It can be a relief to know you're not alone and that you can learn to manage your anxiety in ways that keep it from impairing your life.

GENERALIZED ANXIETY DISORDER

Although there are many types of anxiety disorders, the most common and basic is called generalized anxiety disorder. Generalized anxiety disorder has similar characteristics to the symptoms discussed earlier in this chapter and varies in the spectrum of intensity and magnitude.

It ranges from distinct stress at the low end to potentially debilitating anxiety at the extreme end. All other anxiety disorders—such as panic disorder, obsessive-compulsive disorder (OCD), social anxiety disorder, post-traumatic stress disorder (PTSD) and phobias—can also range from mild to extreme on the spectrum, but the extreme symptoms that result from these conditions can cause very severe life-altering impairments that can sometimes be difficult to treat.

Generalized anxiety disorder is best characterized by persistent worry and preoccupation with an identifiable source of stress or one that doesn't have an obvious explanation. These emotions are often difficult to control or put an end to. Individuals find it hard to stop the mind from racing and can't keep fears and worried thoughts from interfering with their daily lives. They often ruminate and obsess over everyday circumstances related to their jobs, paying bills and their health as well as the health of their families. Many individuals who suffer from generalized anxiety disorder also become overly preoccupied with minor issues such as being late for appointments, forgetting their house keys, disappointing a friend or neighbor. And as I indicated earlier, the intensity of the anxiety and worry is out of proportion compared to the true reality of the stressful situation itself or the possibility of something negative occurring.

One of the important criteria that helps distinguish between whether you're feeling stress or have an anxiety disorder is the duration of the symptoms. According to the *Diagnostic and Statistical Manual of Mental Disorders,* the ongoing worry and preoccupation felt must last for six months or more for it to be classified as an anxiety disorder. Whether you're diagnosed with an anxiety disorder or you're simply a person who is easily affected by stressful situations in life, it's important to get help.

The key to successful anxiety management is to cultivate new ways of coping by implementing the mental ergonomics presented in this book. As you begin to use my tools when you recognize an anxiety trigger, you can gradually build a program of stress or anxiety management that's solid and easy to use. Then you, not anxiety, will choose your response to what life throws at you.

Chapter 2

IDENTIFYING YOUR PERSONAL BELIEF SYSTEM

"It is not the things themselves which trouble us, but the opinions that we have about these things."

—Epictetus

For the most part, ever since you were born, you made sense of the world through your own eyes. And since childhood, you've continually processed and absorbed information that comes from society, school, your parents, religion, the media, your friends and other sources in your environment. This information forms your cemented belief system. We all possess diverse and differing belief systems that represent the internal infrastructure of our lives as we know it. Your personal belief system gives your life meaning and purpose. It typically spans your lifetime, staying with you until you die or until you do something concrete to change it. Your beliefs are carried with you as a fixed template that represents biased truths that you hold on to with certainty.

However, your personal belief system can also create anxiety and keep your mind stuck in rigid thinking patterns.

HOW "FIXED THINKING" CAUSES ANXIETY

Over time, the fixed truths you adopt become the foundation for developing unbending opinions and solid convictions about many aspects of life. The problem is, when you see these subjective opinions as indelible facts, it's hard to recognize that they're just your personal beliefs and that they can actually be changed to a more realistic set of convictions. Your personal belief system can close your mind and blind you to other frames of reference because time has convinced you that you're right or that it's the only way to think. This kind of fixed thinking can cause you to miss out on more rational ways to view situations and enjoy or accept common human experiences.

If you hold on to fixed beliefs about yourself that are negative and/or self-defeating, it will inevitably result in your experiencing ongoing anxiety. For example, if a man has a personal belief or a fixed thought that a "real man" is a strong and stoic individual who should never ask for help because that indicates weakness, he'll be critical of himself whenever he's sad or going through a difficult time. He may also feel guilty for not being able to take charge and find a solution every time one is in need, and subsequently he will suffer alone. He may even feel inferior because he believes, "I should be handling it better." Yet the belief that a "real man" is strong and stoic is simply an opinion, so it's unnecessary to hold it as a standard to compare yourself to. Many men don't feel an intense pressure to be constantly strong and stoic and are perfectly happy and relaxed about whether or not they fit this stereotype. Yet this belief rules many other men and causes great anxiety in some of them. However, it's your choice to accept this or cultivate another personal belief.

Conversely, if your mind is fixed on some sort of positive fantasy of how things are or should be, and someone or something comes along and challenges that fantasy, you can also get scared and anxious. If the belief is strong, you may fight aimlessly to try to hold on to what you think is a basic reality. And, if you're unsuccessful at reaching your fantasy—and you often will be because life is full of variety and constant change—it can create a lot of anxiety.

For example, as a child, a woman may have decided that she'd be married with two kids by the time she turns thirty because it was perhaps what her mother set as her standard. She grew up fantasizing about her dream wedding and two perfect children and often acted out the fantasy with friends, as young girls sometimes do. As she approaches her late twenties, she may begin to get desperate about finding her husband so she can have her wedding and then the children as she planned. The closer she gets to thirty, the more desperate she becomes, which can motivate poor choices and great anxiety. No matter how much her friends reassure her that she's still got time to get married and have a family, her belief that it must be before she turns thirty is so strong that it rules her thinking, and she obsesses about this failure to find a husband.

An ironclad belief system I had years ago was that by age thirty I should have a solid career and my life should have defined goals. Well, things didn't go as planned. I was miserable and felt like I was doing life "wrong." It created a great deal of unrest and worry for me because it was what my family had also assumed I'd do. I felt a lot of pressure because there was a great expectation on me. Plus, most of my close friends had established themselves in careers already. But as I starting working with a therapist on my anxiety, I not only identified my fixed thinking as irrational but I also realized that it was a belief system I created that had no validity. Ironically, the therapy inspired me to become a therapist myself, and I found my calling.

Over time, I had to get comfortable with the fact that it was okay to be thirty and not have it all figured out. I had to accept that I was perhaps simply a late bloomer and not a wishy-washy wanderer with no goals in life. So I became a therapist at age thirty-three. From my work with people who suffer from anxiety and from my personal experience, I've come to understand that you must not only confront your fixed thinking and potentially faulty beliefs, but you must also challenge your own traditional way of thinking. But first, you must understand what belief systems really are. Often this understanding brings clarity and the ability to change.

- **Belief systems** are blindly absolute and uncritically held values and opinions that at times are handed down from generation to generation, such as "people should marry within their own culture or religion." They're made up of guidelines that are based on family values, societal standards and cultural customs. It's typical to not question these beliefs because they were supposedly passed down by sacred ancestors who are assumed to have possessed more wisdom than you. Sometimes, belief systems are even considered by many to have a mythological and almost miraculous quality to them, depending on whom they're attributed to.

- **Belief systems** are also considered to be important ideological principles that possess relative worth. They're ideals and formal assessments of life that are written in indelible ink and speak with an authoritative voice, like "an intimate relationship should only be between a man and a woman." This authoritative voice is so convincing to many cultures that to question certain beliefs is equivalent to committing a heinous crime. You learn them as children and grow up with a strong sense of them being the truth you must live by. But often they don't have to be your truth, especially if they feed your anxious mind. Letting go of these beliefs can be tough if they're deeply ingrained in your mind. But they definitely can be redefined if you're willing to try.

ASH'S STORY

Ash was a fifty-six-year-old man. He was the father of two grown children and loved his family. His parents were first-generation immigrants who were very connected to their homeland culture and its rituals. Ash had a twenty-six-year-old daughter, whom he loved dearly and had a very close relationship with. She had begun seriously dating an American man of a different culture. The man was a successful lawyer who was deeply in love with Ash's daughter. They had begun living together and were planning to marry someday. Ash struggled with a fixed belief that his daughter—whom he did not want to hurt or offend—should not marry outside their culture.

Ash's anxiety had skyrocketed by the time he came to see me because he was torn between enduring the wrath of his extended family for allowing this potential marriage to take place, and the deep love he had for his daughter and the desire for her to be happy. But he was particularly blinded by a self-proclaimed belief that *he* was the one solely responsible for making that decision and making everyone happy. This aspect was freaking him out more than the issue about the difference of culture. He also became anxious because he believed he'd have to disappoint someone because he could not satisfy both his daughter and his extended family. For months he was rattled with anxiety and the potential for unspeakable guilt. He could not sleep and was unable to concentrate on his job. His life was in turmoil.

Over the course of treatment, Ash recognized that he was blinded by the fixed belief that *he* had to play the matchmaker and the judge. It was too much pressure for him to handle. He felt trapped. Finally, after many sessions together, he came to the conclusion that he loved his daughter too much to hold on to his fixed belief and ultimately gave up having to make any decision at all. He concluded that his daughter was old enough to make her own choices and the family would have to live with it. He realized that he alone contributed to his anxiety by taking on too much blind responsibility. Ash identified his belief system around marriage and being a governing patriarch and redefined it. After this decision, his anxiety lessened and his quality of life improved immediately.

UNDERSTANDING AUTOMATIC THOUGHTS

Automatic thoughts, which are fundamental products of your belief system, greatly contribute to causing anxiety. These are responses and reactions to the world around you that are like reflexes you do from habit, without thinking. Since habitual behavior is a natural part of your makeup, your brain is in knee-jerk mode for most of the day. While in this mode, you unconsciously respond to certain stimuli in spontaneous and impulsive ways without even knowing it. This is especially common when you respond automatically when something scares you or you begin to feel anxious. It could be said that we're like walking impulses operating strictly on instinct. The instincts, of course, are guided by your personal beliefs.

The automatic thoughts you acquire over time become part of an internal dialogue you have with yourself. Through personal interpretation, this inner discourse gives each stimulus or event in your life an automatic label followed by an unsubstantiated conclusion—a belief about the probable outcome or why you should feel anxious about it. These labels and conclusions are usually believed, even if there's no evidence that they're right and there's been no analysis of how true they really are. Throughout the book I'll discuss how to change or let go of these thoughts. For now I want to help you understand what they are and how they contribute to anxiety.

Automatic thoughts are beliefs learned from childhood.

You weren't born with your belief system and automatic thoughts already in place for you to live by. They were learned and quickly adopted from various sources during your childhood and adolescent years. You learned your automatic thoughts from society, your parents, your religion, the culture you were brought up in, the media, in school, from friends and other influences. These beliefs might have been learned in response to how you were treated in specific circumstances. Often they're developed as a means to protect yourself or because it was what someone important to you expected.

For instance, if your partner demands that you take on a traditional role in the relationship, even though it may not be what you want, your early years may have taught you that your only choice to avoid conflict is to go along with it. In another example, if you were punished as a child for not getting perfect grades in school or not doing other things that lived up to your parents' expectations, you may still be striving for perfection in your life today. When you make a mistake at work or let a friend down, you may get anxious waiting to suffer the consequences of not being perfect. Even if nothing bad happens, it won't lessen the defensive feelings you'll get the next time you do something less than perfectly—unless you increase your consciousness of the automatic thought that makes you believe that something bad will happen if you're not perfect. The good news is that whatever beliefs you learn in life can be unlearned once you decide to let go of them.

Automatic thoughts are impulsive, as if operating unconsciously.

These responses occur spontaneously, without your even knowing it, and sometimes speak for you before you've had time to consider what's going on and how you really feel about it. Speaking impulsively often leads to regrets, which creates even more anxiety. Let's say you *do* cling to a personal belief that to succeed in your career, you must perform perfectly all the time. If not, you're a failure. One day your boss at work gives you constructive criticism about your recent mediocre performance on a project he asked you to work on. You react negatively because you feel threatened even though he likes you and enjoys working with you. You say things you regret in your desperate defense to cover up your misgivings. Or, conversely, you cower to the criticism and shut down for fear of consequences. But you later bad-mouth your boss to others as a way to retaliate.

In both circumstances, your mind races with catastrophic, negative scenarios about getting written up poorly at the next employee evaluation meeting or even getting fired. Your automatic thought prevailed because your belief is that any criticism is the end of the world and you failed to be perfect. But you're actually not a walking impulse like an animal in the jungle that operates on instinct and natural reflex. You can rise above the impulsive thoughts and automatic urges and think your life through. Your boss might have said those things because he wants to help you advance to your potential. But you won't see that if your beliefs make you automatically jump to negative conclusions. If you tend to react impulsively, rising above your nature to do it is a vital part of retraining your anxious mind.

Automatic thoughts are believed as fact and are rarely challenged.

You may believe that your automatic thoughts are the ultimate truth because they occur so quickly, and your knee-jerk reactions may be so cemented in your brain that you don't ever bother to question them. When you're stuck in a fixed way of thinking, it's very hard to consider other frames of reference because of the blind conviction that your personal belief system can possess. If you are positive that these beliefs are true, you likely wouldn't think to challenge them or even notice what you're doing as you react to anxiety-provoking situations.

When I was a teenager living in New York City, I got it into my head that because I was not very academic, I was unintelligent. All through high school, I failed many classes despite tutors, special remedial classes and many Saturday morning study halls. The sad part was that I believed the automatic thought that I was not a bright child. I never challenged this notion because I didn't know that I could and because my father—who I listened to blindly—often told me that I'd never survive in the world without being a good student. That became fixed in my thinking. And it scared me that I wouldn't be able to survive in the world since I didn't do well in school.

It wasn't until later in my twenties that I was able to open up some space around the false fact that I was unintelligent. I learned that I was actually a relatively smart person who was capable at doing many things—I was a good writer, a good musician and excellent at sports. I stopped listening to the automatic thought that I was stupid.

Automatic thoughts are irrational and imbalanced.

Your automatic thoughts can be reckless and often possess catastrophic characteristics that are usually imbalanced and out of proportion to the severity of the situation. They have mentally distorted traits that can cause you to be very emotional and reactive, like only seeing a situation in all-or-nothing terms. There are very few gray areas in these imbalanced thoughts because the mind is thinking only in extremes.

A classic example of an automatic thought that tends to be irrational and imbalanced is experienced by many of us after a relationship breakup. No matter how mature or independent we are, after a breakup or divorce, many of us suffer and lament, "I will always be alone now" or "I will never find someone else who will love me like he/she did." We are so hurt and scared about being thrust into the single life again that our mind reacts and we have all-or-nothing thinking. The breakup is then blown completely out of proportion and the automatic thoughts run amok. Awareness of how these irrational thoughts don't reflect what your options really are helps you take steps to change them. When you end a relationship, it can be hard to see the options of a new beginning instead of a sad ending because the fearful automatic thoughts aren't being challenged.

Automatic thoughts are reflections of core issues that have been with you for a long time.

It's common to think automatic, recurring thoughts about yourself that are negative and self-critical such as, *I'm so stupid*, or *I should have known better* or *I'll always be a failure*. These thoughts are representations of personal core issues that you hold on to about who you are. They're part of the self-defeating soundtrack in your head that says you don't deserve to get your desires met or to succeed at what you want. This soundtrack is used to define yourself and can be traced back over many years. These core automatic thoughts become your "organizing principles" in life—an often indelible personal belief system template that's formed early and negatively influences your life on a day-to-day basis.

An example of a core issue automatic thought is, *I'm not good enough*. If, as a child, you felt alienated from your classmates and weren't part of the popular crowd at school, you may have labeled yourself as having something wrong with you and therefore being undeserving of friendship and love. This defectiveness becomes an emotional template that's hard to shake off—a principle that you can't break out of. So, as an adult, if you're invited to a party to meet new people, you may have an automatic thought that believes, *No one will talk to me and no one will like me*. It can lead to, *I'll feel awkward at the party and feel super self-conscious. That will make me really anxious.* Because of this, your automatic thought wins, you don't go to the party and the self-defeating behavior continues.

But these self-defeating soundtracks can be rewritten once you face them. The first step is learning to identify them as automatic thoughts, which are part of your old belief system. For example, I had a self-defeating soundtrack for many years in which I told myself that my anxiety caused me to have limitations in my life, and it meant that I was inferior. Why couldn't I be like everyone else? It could only mean one thing: *Something was seriously wrong with me.* This core issue affected almost every aspect of my life—from going to college, to applying for jobs, to dating women. The shame factor had tentacles that reached out to everything that I came across in life. The shame was so toxic that no matter what I did, I still felt inferior to others. The core issue of shame started when I was a child and stayed with me for many years. It was the bedrock of my anxiety.

IDENTIFYING YOUR PERSONAL BELIEF SYSTEM

An excellent way to begin to identify your own belief system and see what your automatic responses are to certain subjects is to complete the exercise that follows. You might want to do this and other exercises in the book in a notebook or a document on your computer that's specifically created for it. Write your own personal beliefs and your family's beliefs about the following terms. While doing this exercise it's important to only write down the first thing that comes to mind. Your recorded response must be your initial, automatic thought. Don't think about it or try to record an answer that will sound "good" or "right." The first thought you have will be your raw personal belief.

An example of a political belief is, *All politicians are crooks*. An example of a career belief is, *If I am not successful, people will not like me*. Be honest. This is only for you. For each of these topics, write down your first thought:

Career

My personal belief is

My family's belief is

Money

My personal belief is

My family's belief is

continued on page 30

MARRIAGE

My personal belief is

My family's belief is

RELIGION

My personal belief is

My family's belief is

POLITICS

My personal belief is

My family's belief is

Are you surprised by any of your responses? Did your beliefs differ from or mirror the way your family saw these subjects? You might find that if you reflect on your automatic responses, your beliefs might have changed over the years. Or you might discover that you have the same convictions you had many years ago and there's very little evidence of change. Either way, the automatic responses you wrote down represent the infrastructure of your current reality or the world as you know it. Or it's the world as you desire to know it or think it should be.

Now I'll guide you through another exercise. Try to remember how you felt the last time someone's personal beliefs about the subjects in the preceding exercise differed from your own. How did that make you feel? Did you label them as wrong for thinking differently? If you can't remember any incident like this, try to pay more attention when it happens in the future and ask yourself those questions. By becoming aware of them, you have a better chance of letting go of those personal beliefs that contribute to keeping your mind anxious.

Let's say you have fixed, automatic thoughts related to everyday subjects like marriage, politics or religion and you have a friend or a close relative who does, too. In fact, you might feel so convicted about your automatic thoughts that some of them even define your personal value system. Perhaps these are even values and beliefs you have never questioned or challenged in the past because they feel so "right" to you. They feel SO right that they appear like absolute truths. But during a conversation about the topic, you discover that your beliefs are opposite to those of who you are talking to. How might you react?

Because beliefs can be very powerful, you may actually be shocked that other people could think the way they do. How could they believe that? They must be crazy! You may become suspicious of them for thinking in ways that seem so illogical to you and, more important, for thinking so acutely in an opposite direction from you. You may even form quick judgments that determine them to be unstable or untrustworthy people. The irony here is that these people, who you may jump to label as unfit to walk the planet, might be thinking the same about you. Why? Because, like you, they could also be stuck in their own beliefs and unable to see beyond them.

Now think about how you've felt in the past when you encountered someone whose beliefs were similar to yours or imagine how you'd feel if this happens in the future. This time, you'll most likely develop respect for these people almost automatically and trust them almost implicitly based on the shared belief. Labels like "unstable" or "untrustworthy" will probably not come to mind, even if they actually are! Your automatic thoughts will probably create more positive judgments of them based solely on having a common belief. You may quickly give them labels such as:

- He/she is a very good person.
- I'm sure this is a very nice man/woman.
- He/she must be intelligent to have that same belief.
- He/she has a good head on his/her shoulders, like me.

These kinds of automatic judgments are based on a shared belief. Yet you may barely know the person. This kind of response can cause you to judge a person incorrectly. It's important to recognize how these automatic thoughts based on beliefs that were set in stone many years ago can play out in your life now. They can be a catalyst for anxiety and keep you from controlling it. I'll be referring to them in other chapters. For now, begin to increase your awareness of your personal belief system and the kinds of thoughts that you automatically go to. Imagine that you are changing the personal "settings" of your beliefs, as you would on your computer, by unchecking the little box of your former default system. It's okay for people to think differently than you do, and recognizing this is a good step toward retraining your anxious mind.

Chapter 3

CHALLENGING
THE NOTION OF
"CONSENSUS REALITY"

---

*"Reality is merely an illusion,
albeit a very persistent one."*

—Albert Einstein

Over the many years that I've been treating patients who suffer from stress, excessive worry and even severe anxiety disorders, I've discovered that many of these individuals cling to the idea that there's a consensus reality—one way to think about and respond to something specific based on what they've been taught is believed by most people. Consensus reality is a mental view that we unknowingly subscribe to as the ultimate truth. It's a narrow and limiting way of looking at life that accepts, sometimes with fervor, that a single, unified reality in the world for everything exists, and that you must abide by it. This kind of reality is a fictional conviction that often possesses inflexible rules about how to think and act. According to this concept, individuals believe that there's only one opinion or one path to follow for how to succeed in life, how to love one another, how to worship God, or what and how to feel about things that come up from day-to-day living.

For example, throughout most of my childhood and young adulthood, I clung to a consensus reality that I picked up from my father that said, *You should never trust anyone.* My father had apparently been treated poorly by his family of origin (whom he hated) and then later by many business partners throughout his life, so his motto was, *Love everyone, but trust no one.* He was very bitter about it and often gave me vivid examples of the many betrayals he'd suffered in his life. Sadly, he never connected these events with how profoundly hurt he was. He was only tuned in to the anger he felt about it. Because the anger clouded his vision, my father did not allow himself to heal from the ways he'd been burned because he never challenged his thinking, so instead he decided that all people were bad. His consensus reality that *No one should ever be trusted* became his safety shield. He then passed the same absolute belief on to me. He meant well by trying to protect me, but it affected me deeply. Because I adopted his reality, I turned into a suspicious person who feared letting others get close to me, like my father did. The one-way reality that I never challenged left me lonely and scared for many years.

This kind of rigid thinking contributes to increased anxiety over a prolonged period of time and can leave you scared and ambivalent about people and what's going on in your life. You may spend inordinate amounts of time second-guessing about what's right or wrong and even wasting time trying to be perfect. Or it can cause you to feel oppressed or inhibited by the inflexible regulations that you are convinced you should adhere to. Either way, you can become a slave to the cues you get from what society seems to dictate, or family expectations of you, or religious mandates or other people or areas of your life that matter to you. By thinking this way, you deprive yourself of being able to focus on your internal voice or, better yet, your own view of your life and the people and circumstances you encounter.

I'll discuss how to create your own reality later in the book. Right now I want you to understand where your current perspective of reality may come from. Awareness is the first step to challenge any consensus reality way of thinking.

It's critical to understand that there is no such thing as a *real* consensus reality—and there is no such thing as reality at all. In fact, there is just how you see life through your own personal filter. Consequently, you have the power to change the filter and develop a new vision of life. Once you can do that, a whole new world of conscious thinking can open up for you. In time you may begin to comprehend: *I have the power to create my own reality.*

If you recognize that you do indeed have the power to create your own reality, you won't be a slave to anyone's criteria of how to be a man or a woman, or any other imposed opinions and/or judgments of yourself and how you should act or respond to various situations. At that point you've liberated yourself from dependence on external cues for validation and acceptance. You'll become free to develop your own internal evaluating system based on *your* criteria for being a human being, based on a reality that you create for *yourself.* This very mindset can easily begin the process of reducing the frequency and severity of anxiety symptoms.

DREW'S STORY

Drew worked as an attorney in the entertainment industry in Los Angeles. He was married with no children and came to therapy seeking help for what he called his "freeway phobia." He reported that he had no other problems in his life and insisted that his life was great, except for the driving problem. Drew said that he was able to drive on surface streets and country roads, but he could not get on a freeway without experiencing severe anxiety symptoms. Whenever he tried to enter the freeway, he'd experience debilitating issues, such as heart palpitations, labored breathing, sweaty palms, dizziness and choking sensations. It hampered his life and his ability to get where he had to go quickly.

Drew admitted that he had always considered himself to be a mildly anxious person, but in the last two years, his anxiety had mushroomed into an anxiety-producing freeway phobia. It all started one day while driving home, on the freeway, from a stressful day at work. Out of the blue, he suffered a massive panic attack that almost caused him to pass out. He was so scared by this that after the symptoms subsided, he drove himself to the emergency room at the nearest hospital. They told him he was fine after various tests were run. There was no visible reason for his symptoms, and he was sent home with no formal diagnosis. That added to his anxiety.

Over the next few months as therapy continued, Drew began to discuss some of the high standards he held for himself and also how often he was very self-critical. He acknowledged that he was a people pleaser and wanted to make everyone happy all the time. He had been raised in a family where pretentiousness and always looking good, feeling happy and being successful were very important to his parents. From a young age, the pressure was on him to live up to his family's standards and to follow in the footsteps of two seemingly perfect older brothers. As adults, one became a successful criminal lawyer, and Drew looked up to him, viewing him as a model for the way a man should be. His other brother was the head of a prominent movie studio in Hollywood. They were hard acts to follow.

During therapy Drew also shared that his work environment was very competitive, and the need to produce favorable results on a regular basis added to his pressure. He felt a strong need to achieve according to his family standards, and it stressed him greatly. He also admitted that at one time he had considered leaving the field of law to become a high school teacher—a dream he always had. Drew believed that was his true calling, but he was afraid of what his family and friends would think of him if he quit his job to teach, especially his mother. He believed they'd look down on him for not having a respectable and distinguished career. He also admitted that sometimes he thought that he probably became a lawyer just to please his parents, who were also very accomplished. His father was a retired heart surgeon and his mother was an engineer. He grew up feeling inferior to them and deep down believed he could never measure up to their level of success, no matter what he did.

Another important aspect of Drew's life was that, since childhood, he considered himself to be very close to his maternal grandmother, who was a Holocaust survivor and the dominant matriarch of the family. He said she was more like a mother to him than his real one. Over the years she had been very open with him about what she had suffered through. Although he always felt uncomfortable about listening to her stories, Drew felt obligated to be there for her at all times. He could never say no to her, and, in time, serving as her private support system became his exclusive responsibility. He tried to be very careful to never add any additional pain to her life, so he learned how to walk on eggshells around her. He later shared that because of this relationship, he began to walk on eggshells around everyone. Doing what he could to please people and not say or do anything that could annoy someone seemed like a safer way to be.

Drew came to therapy to get help for his freeway phobia, but in actuality, he discovered that it wasn't the freeway that was stressing him out. He himself was the cause of his escalating anxiety. He realized that he was going along with a consensus reality learned at a young age that was very rigid and unforgiving about career, success, relationships and family obligation. He felt trapped in the perceived reality that his family had unwittingly imposed on him that had no wiggle-room whatsoever to think or be different from it. But most important, he felt terrified that he would lose his family's love if he ever strayed from what was expected of him. Drew allowed this consensus reality to take over his thinking by adhering to the following inner dialogue:

- *To be accepted in my family and society, I must always be successful at everything I do.*

- *To be accepted in my family and society, I must have a prestigious career.*

- *To be accepted in my family, I must always consider the needs of the family first.*

- *To be accepted in my family, I should never disappoint anyone.*

- *To be accepted in my family, I should feel responsible for everyone's feelings at all times.*

- *To be accepted in my family, I must make sure grandma never feels any pain.*

Once Drew took this first step to identify that he grew up learning these hard-and-fast rules for being accepted, he was able to do the work to let go of the consensus reality that had dominated his life until now. Acknowledging it was the first, critical step. Until you see where your problem originates, it's hard to do something positive about it. Once Drew acknowledged the beliefs that drove him, he was able to claim his life on his terms and temper the stress that caused his freeway phobia. He's still working on breaking the habits of responding with his consensus reality mindset that he was convinced was his only way. But he's much happier as he slowly creates his own reality—and he's looking into the possibility of teaching at least part-time.

THE TYRANNY OF THE "ABSOLUTES"

Another excellent way to expose any consensus reality you may have and to identify the times you may cling to it is by listening to the absolute language you use in conversation or self-talk. For example, think about when you see words such as *should, shouldn't, never, always, everyone, no one, everything, nothing, must* or *ought*. These absolutes or blanket words can cause you to inadvertently go along with the idea of the mono-reality—one absolute way of thinking that leaves no room for exceptions or alternatives—by using words that limit your thinking. They also can make you feel pressure to always live up to them, or they may contradict what you actually want to do, which creates conflict about what you *should* do. The most destructive absolute words of all are *should* and *shouldn't*.

Think about it. If you say to yourself, *I should have known this would happen,* or *I should be more productive,* you imply that there's an invisible manual or instruction book floating around out there in the air with absolute decrees that dictate how to be a good human being, and that there are rules that must be followed and you're not getting them right. Therefore, you'll probably convince yourself to feel guilty. This kind of guilt can subsequently lead you to feel ashamed and powerless.

The dangerous "absolute" quality of the word *should* also implies that you could potentially have access to a crystal ball and can predict the future or have some kind of mind-reading device that can magically reveal what other people are thinking. The word *should* is one of the most irrational words you can use.

For a long time, I clung to the tyranny of the "should" in making decisions. The unbending "absolute" here was, *I should never make any mistakes.* I was so terrified to make wrong decisions about even small things that I ended up making very few decisions at all. I had to rely on others to make them for me or sweat bullets when I had to make them on my own. I couldn't see that I naively believed there was always a right decision out there to be made. So I had to try my best to use the proverbial crystal ball and look into the future to make sure I chose wisely. Naturally, this set me up for years of anxiety because we all know that we cannot predict the future. My one-way reality from using the "should" absolute did not leave me any room for error.

I encourage you to try to replace the word *should* with the phrase, *I would prefer*. For instance, if you're thinking, *I should be more productive*, when you feel you *should* get more done, you can replace that with a more rational thought like, *I would prefer to be more productive with my life. Let me see what I can do to change how I work so I can accomplish more.* This thought takes the guilt out of the equation and may in fact empower you to take action instead of remaining passive. That's a good reason to eliminate *should* and *shouldn't* from your vocabulary whenever possible!

I remember treating a patient a few years ago who often used another strong absolute—the word *never*—in many of her sentences. She especially used it when she thought of her future. She wanted very much to get married and have children. So whenever she felt lonely or a relationship with a boyfriend ended, she would think, *I will never meet the right man* or *I will never have a family*. Sometimes her absolutes changed to, *I will always be alone*. At age thirty-eight, Bethany didn't realize that although she did have a biological clock ticking in the back of her mind, the use of the word *never* was causing her such epic anxiety that she could not function at work and tend to her responsibilities. The word *never* is irrational in its nature because *never* is a future-based absolute that can't be substantiated. But more important, it was connected to her consensus reality that she picked up from society that a woman is only defined as a woman if she has children.

Another consensus reality she clung to was that you can never be happy unless you experience the joys of raising kids. These two concepts were freaking her out because she never had any reason to challenge them. In our work together, I helped her to distinguish between other people's reality of happiness and their concept of what it is to be a woman, and what she felt comfortable believing herself. Once she started to develop her own voice about these two subjects and created her own reality, her anxiety decreased. Bethany realized she still wanted to have children and believed it would make her happy, but it was no longer a desperate search based on someone else's reality. She then felt less pressure to pursue her dreams because now they were on her terms.

A decisive aspect of retraining the anxious mind rests on the premise that your personal views of yourself and your environment are the major determinants of how you feel and act. By clinging to personal beliefs about yourself and the world around you, you hold them tightly as the ultimate truths you adhere to and unwittingly embrace these truths as if they were factual, rock-bottom data. However, in time, you can break the cycle by acquiring the mantra used by many self-empowerment leaders: "Don't believe everything you think." This means that when you're anxious, don't immediately trust your automatic thoughts because oftentimes these thoughts are irrational. Thoughts are not facts. They're just thoughts and sometimes don't need to be given so much importance.

For example, if early in your teenage years you never felt like you fit in with your classmates or were teased, over time you may isolate yourself from others, leading to feeling as if you didn't belong. This can lead you to create a negative opinion about yourself—that you're inadequate or worthless. From there you may formulate other solid, believable negative beliefs about yourself that gradually become perceived as fact or the rock-bottom data. These negative beliefs or mental distortions could generate thoughts like, *I'll never have friends again, I will be alone forever* or *I'm not worthy enough for people to like me, so there's no reason to bother to try to interact with anyone.*

Consequently, you'll unconsciously begin to write a story, a screenplay almost, written in indelible ink, of who you are, based on the inaccurate belief that your mind has accepted: *I don't belong.* Years may pass and you'll continue to cling to this story, not even realizing that you made it up or that it's simply a distortion you naively adhere to based on a situation from many years ago. Then, you ultimately begin to mold an inner core consensus reality, an unyielding conclusion about what you are, which may be something very irrational like, *I'll always be an inferior human being.* Unless you let go of these beliefs, they will continue to keep your mind anxious.

TOOLS FOR RECOGNIZING AND CHALLENGING YOUR IDEA OF "CONSENSUS REALITY"

Once you become aware of any consensus reality that helps to form the foundation of your personal belief system, you can begin to change your reality to one that's more realistic. In Chapter 7 I'll discuss in depth how you can create your own reality. But first I want to further help you recognize the absolute beliefs you have about how you and those around you should behave. The beliefs in your current reality may have begun when you were very young. It can take time to identify them and acknowledge that they don't need to apply to you. But once you do, you have another piece of the power kit you need to retrain your anxious mind.

For now, start to pay attention to thoughts that are based on beliefs you learned from your family, religion, in school, from friends or even from a romantic partner. Since awareness begins the process of retraining your mind, it's important to really pay attention to any absolute thoughts you may use to view what's going on in your life. To accomplish this, ask yourself the following questions, and try to be honest with your answers:

- **Am I living with the idea of a mono-reality?** If you judge people and circumstances according to beliefs you've been accustomed to using as the standard for how people should live, it limits what you find acceptable. This can lead to an attitude of "my way or the highway," which alienates people. And when something conflicts with your beliefs or someone challenges them it can increase anxiety.

- **Am I closed off to alternative realities/possibilities and points of view?** When your thinking is rigid, it can put a lot of pressure on you to keep your world working the way you think it should since many people will disagree with you. Not being open to what other people think can create stressful static between you and those you have relationships with, whether at work or in your personal life. It's hard to get along with people when you refuse to consider other points of view.

- **Am I focusing on external cues for guidance and validation?** When you need approval from other people to feel like you're doing well, it's hard to have good self-esteem. That also keeps you locked into their values and beliefs so you don't have freedom to develop your own. You might not be fully comfortable with what you've been conditioned to believe, but trying to think for yourself can seem scary.

That can create a conflict inside of you between needing to please others and wanting to please yourself.

- **Am I focused on old, self-imposed or family-imposed rules about how to live my life?** Decide if you really believe what you were taught or if you just go along with the beliefs to fit in or because you're scared to think for yourself. Think about how these beliefs came to you. Were they drilled into your head and you adopted them to avoid problems that may have come up if you stepped outside of them? Or did you adopt them out of fear—to keep your life structured in what you thought was an acceptable way? It's time to decide how you want your world to really be.

- **Am I limiting my experiences in life by using absolute words such as *should, never* and *always*?** If you believe you "shouldn't" do something, you probably won't do it, even if you'd like to. If you live with many "shoulds," you can get stuck doing things you don't want to do. Believing that you must "always" do things perfectly sets yourself up to be let down if you fall short. Those absolute kinds of words can keep you in a box with invisible walls that block you from stepping out of your comfort zone to do things that would enhance your life. Trying to live up to these absolutes can be a great source of anxiety since you're human and won't always be able to uphold them.

REPLACEMENT THOUGHTS—
THE FIVE-MINUTE RULE

After you recognize any ideas of a "consensus reality" that you are clinging to as rules you must live by, give yourself five minutes to step back and try to respond differently. Think about what these beliefs really mean to you and if you want to stay locked in the rigid thinking that's contributed to your anxiety. It's important for you to take your blinders off so you can see the variety of options you have for how you can perceive someone or a situation. This is a process of unlearning the old habits of automatically responding the way you always have. Begin to expand your horizons by regularly replacing your "consensus reality" thoughts with the following affirmations:

- **I will remind myself on a daily basis that there is no such thing as a one-way reality.** Holding on to a belief that there's only one way to think about what happens in your life is your choice, not the way it has to be. It takes time, and you often must consciously remind yourself to let go of the habit of going into one-way thinking patterns. As you practice and get used to it, broadening your perspective will become easier.

- **I will open up my mind and consider alternative realities and points of view that I may not have ever seen before.** They're out there if you look. Ask yourself, "What are my choices? What might be a better way to view this?" When people offer a perspective that's different from yours, ask them to explain their reasoning instead of just negating it automatically. When you hold on to a consensus reality, you may not listen to someone who thinks differently. But asking about that person's alternate views can enlighten you to new ways of thinking.

- **I will practice letting go of my dependence on external cues for guidance and validation.** People pleasers look to others for approval. It can be stressful to always need others to validate how you *should* think and *should* view a situation that arises. When you remove your blinders and accept that there are other reasonable perspectives besides the ones ingrained in you, your anxiety can decrease. You'll begin to develop a true sense of self as you gradually learn to think for yourself and seek self-approval rather than only trying to attain it from others.

- **I will consider focusing on my own reality that I create based on my criteria for how to be a human being.** When you begin to consider "What do I really think based on logic and the facts?", you'll find it easier to look past the reality you've been living in to see a better one beyond it. Opening your mind can reduce the triggers that create anxiety.

- **I will replace absolute words, such as *should, never* and *always,* with more balanced and realistic language.** It's important to become aware of how you use these words so you can find alternatives that don't put pressure on you. The more balanced you can get in your thinking, the less anxiety you'll have from it.

Chapter 4

BALANCING THE DUALISTIC MIND

---

*"Looking out into the universe at night,
we make no comparisons between
right and wrong stars, nor between well
and badly arranged constellations."*

—Alan Watts

Do you tend to see yourself and your life in black-and-white terms—that everything is either one extreme or the other? Viewing yourself and your world like this is another way you trigger more anxiety in your life. Thoughts get distorted when you think with what's referred to as a dualistic mind. This means you label everything in your life in only one of two ways: for instance, either or extreme, like "I'm the best student in school" or the other extreme, like "I am a complete failure." This may seem like a good way to create a sense of security and control over life's uncertainties, but a dualistic mind in fact tricks you into believing you have life figured out. But you usually don't—who does?

This false sense of security allows you to feel that your life is spelled out for you and it's not necessary to struggle and search anymore. That can feel good in a world where things are constantly changing and people make a great effort to just survive. But the feeling is only temporary, and when reality hits, it can leave you more anxious than you were before.

BLACK-AND-WHITE THINKING

While it might feel good in the moment, thinking with a dualistic mindset creates an all-or-nothing mentality that narrows your vision and triggers insecurity in the long run. It also instills more rigidity and conspires with the consensus reality discussed in Chapter 3 to view your world and perception of situations in ways that create extreme judgments. For example, a dualistic mindset compels you to judge yourself as either:

- Right or wrong
- Good or bad
- Strong or weak
- Smart or stupid
- A success or a failure

An unflinching dualistic mind has no balance in its thought process. It's all one-sided and usually very severe and stubborn. This mindset distorts how you see yourself, and if you're not perfect it automatically casts you in a negative light. Not being as smart as you'd like does not make you stupid but a dualistic mindset says it does. And you're not a failure if you're striving for success but haven't reached it in your self-imposed time frame, but a dualistic mindset says you are. Those kinds of thoughts keep you from seeing all of the positive options that exist in between the two extremes and greatly limits what you think you're capable of doing.

I know all too well how extreme thinking not only can create anxiety but can also block your ability to enjoy the pleasure that can be found between the black-and-white. The memory of being a passionate competitor in athletics when I was younger and the anxiety caused by wanting to be the best at everything I did still lingers. There was no room for a mediocre performance or a less-than-perfect day. Whenever I played tennis, softball or basketball with friends, my dualistic mindset became my worst enemy. If I batted four times during a softball game, got three base hits and was out only one time, that one out meant it wasn't good enough for me to feel a sense of accomplishment.

After the game, I'd sadly fixate on the one time I was out and let that define my whole performance at that game. I'd run it in my head, over and over, for the rest of the day, filtering out the three hits I got. My focus was entirely on the one out, and my dualistic mindset declared me a loser because I didn't hit a perfect four for four. I was either a great player with a perfect day of getting a hit every time, or, because of one out, I was a lousy player or a loser who was unworthy to show his face on the field again. I could never celebrate getting three hits because of this, and as you might imagine, playing ball became an anxious time for me.

The fact is, life doesn't work in that one-sided way. It's full of subtle balance and varying degrees in every area of being human. There are very few situations that are not. Remembering this can counter tendencies to think in terms of absolutes. Keep these two principles close at hand:

- *All circumstance is neutral.*
- Nothing is set in stone because virtually *everything is negotiable.*

The approach to life that these beliefs give you is valid in most cases. However, keep in mind that there are a few exceptions involved with these concepts because the phrases *all circumstance is neutral* and *everything is negotiable* are in themselves absolutes. So when you take into account these principles, just remember that they do *not* include willfully and intentionally hurting or causing harm to anyone in any way, shape or form or breaking any laws. Basic moral codes are *not* negotiable and they are *not* neutral circumstances. This is a critical area where the line is drawn and there's no gray.

ZACH'S STORY

Zach was a thirty-two-year-old graphic designer who sought treatment for his symptoms of acute anxiety and excessive worry. He came to me a year after moving from New York to Los Angeles seeking a better quality of life. He admitted that the transition was difficult. Finding full-time work had become more of a struggle than he expected. He had also recently married his long-time girlfriend, Donna, who worked part-time as a waitress at night. The two were happy together on a personal level, but their financial situation was bad. They were having such a hard time making ends meet that Zach had to ask his parents for money until he found a job.

Adding to the stress, Zach and Donna talked about having children soon. Zach was under tremendous pressure to make things work. He felt like a failure as a husband and believed he let his wife down by not providing for her by getting a good job soon after they moved. He also felt like a disappointment to his parents. He acknowledged that moving to Los Angeles was his idea and since he was unable to find work, he felt shame and regret about his decision. Zach told me, "I should have thought it through better." He was also scared that Donna would leave him to find a better provider if things didn't change soon.

Zach confessed that he'd always been a perfectionist. I heard stress build in his voice when he explained how much he hated to fail at anything and how his standards and expectations were very high, much more so than that of his friends. He remembered being this way all the way back to grade school. As a little boy, Zach never quite grasped the basics of math and algebra and often needed a tutor to complete his homework. His classmates had an easier time, so Zach gradually convinced himself that he was dumb and that something was wrong with him. The fact that he did well in most other subjects, sometimes better than most of his classmates, was a moot point. If he didn't excel in everything, in his mind he was stupid.

One day an impatient teacher, insensitive to Zach's struggles with being perfect, humiliated Zach in front of his whole class. After that incident and countless others, he began to feel scared of any teacher calling on him or being asked to solve math problems on the black-board in front of his classmates. He also pretended to be sick on the days he was supposed to attend math class and he missed a lot of school because of his aversion. After finishing grade school, he eventually forgot about the experience and went on with his life. Or at least he thought he did. But his *I must be stupid* belief stayed with him long after.

While you may not often think about incidents that deeply affected you long ago, as with Zach, they stay with you, silently waiting for an incident to bring them up. As an adult, Zach achieved a reasonable amount of success through his work as a graphic artist, and Donna loved him unconditionally. But he was still haunted by the image of being discovered as the "dumb guy." He felt as inadequate as a man as he did as a schoolboy who struggled with math. That old, painful memory felt as real today as it did when he was younger. While Zach came to therapy seeking help for his symptoms of anxiety due to his financial hardship and other things he worried about in his life at that time, in treatment he recognized that what he felt went much deeper than his current situation.

Zach saw that his need for perfection at all costs was the root of his profound distress. He was never able to master solving math problems, yet subconsciously he used the kind of exactness and discipline that's required for doing math to achieve in other areas of his life to make up for his shortcomings. That motivated him to tackle everything with harshness, self-criticism and unbending rules. Over the years, Zach developed an all-or-nothing mentality about every aspect of his life in order to overcompensate for what he thought was a deficit of character and basic lack of intelligence.

Zach's dualistic thinking led him to believe that he had to be perfect when making all decisions about his career, in addition to being perfect in his roles as a husband and son. Any deviation from perfection was labeled as wrong or a complete failure in his mind. He couldn't see himself as a balanced human being with both strengths and shortcomings. Most of his thoughts and actions, and sometimes even his emotions, were judged as acceptable or unacceptable, as right or wrong. And usually he felt wrong. In Zach's life, no circumstance was neutral and very little was negotiable. His high standards for perfection made most of his self-judgments land on the negative side. Zach fell victim to his dualistic mindset by adhering to the following inner dialogue:

- *To have value as a person, I should never fail at anything.*
- *To be liked and admired by others, I must never make a mistake.*
- *To be seen as an intelligent person, I must be good at everything.*
- *If I am perfect at everything I do, people will not abandon me.*
- *All my decisions in life must be well thought out and must be correct ones.*
- *To ask for help means that I'm a failure.*

Once Zach became aware of how he brought his perceived failures from not doing well in math into every area of his life, he was able to use my tools to let go of his rigid perfection standards. It enabled him to slowly release his harsh self-judgments and find a kinder, more accurate way to view himself and his accomplishments. He was also able to relax a lot more with his situation, knowing he was doing his best. That instilled a much more positive state of mind in him. Easing up on his unfair judgment of himself allowed him to focus his energy on his good qualities, which he credits for eventually getting a meaningful, satisfying job that fulfilled his reasons for moving to the West Coast. Zach felt like he was reborn once he was able to get his black-and-white thinking in check. As he changed his way of thinking, a more positive world opened up for him.

FINDING YOUR GOLDILOCKS ZONE

Balancing your dualistic mind opens up a new world of thinking that you may not have known existed. When you're stuck in a fixed view of life, it's very hard to see outside of it. But once you're able to see virtually all circumstances as neutral and that every aspect of life can be negotiable, you can essentially create your own reality. That gives you another opportunity to break down the limiting walls of consensus reality. In a way, allowing this concept to open up in your life is like feeding yourself oxygen since it unlocks a flow of freedom that's life giving. Letting go of dualistic thinking helps you breathe easier and begin to shed your old suffocating ways created by thoughts that limit your life.

A good example that illustrates the importance of generating the balanced gray areas of thinking is our solar system. One reason that planet Earth is able to maintain life and keep us alive is that its location in our solar system in juxtaposition to the sun is perfectly situated in what's called the "Goldilocks Zone." This is the orbital position that produces the temperate and habitable conditions that are "just right" for sustaining life. If the Earth were closer to the sun, liquid water would evaporate because the planet would get too hot. If it were further away from the sun, liquid water would freeze because it would get too cold. Imagine what would happen if the Earth became a dualistic planet and strayed either toward or away from the sun. Life on Earth couldn't survive those kinds of extremes. But somehow, by the grace of a gravitational pull that keeps it balanced in the Goldilocks Zone, the Earth remains in the gray area of its orbit—the good balance that enables our lives to continue in relative comfort.

This analogy is similar for you. When you can create your own Goldilocks Zone—a balanced gray area of thinking—you'll manage your life better. As your thoughts get more in balance, you generate conditions that are more favorable for living well because you're less likely to fall victim to thinking that sends your thoughts into irrational extremes. Finding that balance creates more calm inside you, which feels good and allows for being more rational.

As I said earlier in this chapter, it's human nature to go astray from that zone by entertaining dualistic appraisals of yourself and the world as right or wrong, good or bad, or strong or weak. You need to be vigilant about the kinds of thoughts you create and remain rational by looking for the balance in any situation of life. And guess what? *The balance is always there.* You just have to look for it with an open mind and a willingness to keep your judgments from going to extremes. Remember, like the old adage says, success is 90% of how you handle a situation.

BECOMING MORE REFLECTIVE

Retraining your anxious mind means you rise above your dualistic thoughts by learning to identify your position as rational or irrational, as balanced or imbalanced. In time, you can acquire the healthy coping skill of being more *reflective* than *reactive*. Reflective means you think about what's going on instead of making a snap decision. It also means looking at the situation or judgment more carefully to note the details instead of letting your first reaction take you to an extreme. Being reflective requires choosing to look for the balance, knowing it's there if you want to find it. For example, learning to be more reflective than reactive is like installing a light dimmer mechanism in your mind. When the light is "on," perhaps you are too reactive to many life situations. And, since we *do* need some anxiety to survive in the world and it's almost impossible for the mind to be on the "off" position, a dimmer device is the next best thing.

When you react immediately, your emotions often drive your perception and dictate your response. This usually gives you just enough time to look at both sides instead of what's in between since there's no time to look deeper. Then your view is either good or bad, right or wrong, smart or stupid. Reflecting allows you to first calm down so you can see what's between the extremes. When you do this, you can weigh your options and make a more rational decision than if you just react to what immediately hits you at that moment.

If you're reflective during difficult times and use this new way of thinking, you're less likely to react negatively or wander off into extreme thinking. Only when you're in this new place can you begin to create your own reality. That adds a great tool for you to control your anxiety, since you eliminate some of the habits that create an anxious mind.

Let's look at this closer. If you believe that your life and destiny are, in a sense, pre-ordained and there's no more room for search and discovery, you'll react negatively when things *do* change—and they always do—because your tightly bolted view of life is threatened. You'll become reactive because the right or wrong, the good or bad that define your life and the lives of others in the consensus reality that you adhere to get compromised. At this moment, you can't be reflective and can't find your Goldilocks Zone since you're not even aware that it exists. You only know one way—the one that keeps you stuck in seeing your life through the vision of thinking in extremes.

MYRA'S STORY

Myra, a twenty-eight-year-old woman who worked as an administrative assistant for a large escrow company, came to therapy complaining about symptoms of anxiety and excessive worry about her future. She revealed that she was raised by a very controlling mother who was poor for most of her life. This led to her developing a strong, almost unrealistic work ethic. Myra remembered what her mother told her every day: "Life is not about fun and games. Life is about hard work." Myra also recalled that on many occasions her mother said that a responsible adult is always vigilant about money, work and most of all, people. She taught her to never trust anyone, saying, "Once you let your guard down, people will take advantage of you."

Over the years, Myra unknowingly internalized her mother's words, and they eventually became a part of how she responded to life and formed her beliefs. From age ten, she began to excessively worry—every day—that she would suffer from irreparable consequences and feel great shame if she did not perform at 100% all the time no matter how insignificant the task was, including schoolwork, athletics, chores and her job. Most of all, it meant Myra could never disappoint anyone, even her best friends. She became self-conscious about her looks, clothes and how well she got things done. That led to developing a dependence on other people's opinions for her self-image, resulting in her becoming a people pleaser and a perfectionist.

Then Myra began to police herself. If she came home from work and tried to relax on the couch for a breather, she felt guilty that she wasn't being productive. If she overslept or was late to an appointment, she'd be very critical of herself and experience mild panic attacks. Taking vacations was forbidden because that constituted fun and Myra equated having fun with being negligent. She believed that disaster would be waiting for her when she returned from a break she enjoyed.

Myra's reactivity was directly associated with fixed black-and-white thinking about what a responsible adult was. At work she was either a frantic woman getting very little sleep and operating on all cylinders 24-7, or she was a lazy slacker. And she constantly compared herself to others. She over-monitored herself as interesting or boring, pretty or ugly, or smart or stupid. She could not break out of her extreme thinking and find her Goldilocks Zone. There was no balance in how she assessed herself and how she compared to others in her world.

Over time, Myra learned to balance her dualistic mind and be more reflective than reactive by using the Five-Minute Rule at the end of this chapter to step back and identify the underlying source of her dogmatic inner voice. Then, with practice and being more conscious of her thoughts, she learned to cultivate her own voice—one that was compassionate and more realistic. Slowly Myra developed a temperate voice that benevolently searched for the gray areas in her extreme thinking. She even admitted that she hated feeling scared all the time, and she realized that all she was clinging to was an old script that kept her on a short leash.

Eventually she redefined what a responsible adult was that was based on *her* terms and reframed her view of people based on *her* own philosophy of life. Myra recognized that up until that time she had reacted to situations with a knee-jerk response that she was unaware of. She was convinced that the all-or-nothing standards she lived by were normal and that to survive in the world, she had to cling to them. During her treatment I gave her the tools in the section that follows, and they helped her cultivate her ability to be reflective. She stopped her escalating negative thoughts by gently reminding herself she was no longer a little girl living under her mother's roof and was an independent woman who understood that a responsible adult actually did the opposite of the kinds of things her mother taught her.

Myra now knows that a responsible adult focuses on good self-care and cultivates balance in the gray areas of her life on a daily basis. And a responsible adult lets go of striving for perfectionism and excessive worrying about the future. A responsible adult needs to take vacations and afternoon breaks and requires the love of others, too. Once Myra accepted that, she was able to relax into her new, healthier mindset.

TOOLS FOR BALANCING THE DUALISTIC MIND

You *can* balance your dualistic mindset! What's most important is to recognize that you go to extremes and can slowly begin to find more balance in your thoughts.

It will probably take time to break any tendencies of taking your thoughts in a direction that you've used for a long time. At first, even your expectation for the time frame in which you will stop thinking in black-and-white may begin as an extreme. That's because once you identify this pattern and decide to change it, you may want it to immediately stop. But going from thinking in extremes to seeing and embracing a more middle ground probably won't happen immediately.

Start by accepting that any small change you make is progress. Ask yourself the following questions and let the rationales behind them slowly sink in. These will help you recognize any dualistic tendencies you may have and stoke your awareness so you can let these types of thoughts go:

- **Am I thinking in extreme terms and using black-and-white thinking?** Remember, solely looking at extreme options keeps you from considering all the possibilities that lie in between the black-and-white choices. It denies you many opportunities to find solutions that could ease your anxiety. When you allow yourself to see and use a middle ground, your anxious mind can relax.

- **Am I reacting emotionally when things don't look right?** If you don't achieve the perfection you desire, you might get frustrated and angry with yourself. Judging in the extreme sets you up for more instances to be disappointed. Since your version of right or good or successful has to be at the top extreme for you to be satisfied, anything short of that can make you see yourself as wrong or bad or a failure. Do you beat yourself up if you fail to get something as right as you think it should be? Dualistic thinking creates all those negative emotional feelings. As they build, they lead to more anxiety.

- **Am I looking for too much certainty in a world full of uncertainty?** Do you try to control everything around you in order to have things go the way you think they should? Remember, since you can't control all the outside influences, things will often not go exactly as you want them to. Yet in an effort to go to the extreme you want, you may try to fight the system and control others around you. If that doesn't work, anxiety builds. When you can't be satisfied with less than the extreme you want, it can push you to keep trying unsuccessfully to make the certainty you want happen. You may be seeking perfection, which is impossible to attain. Yet your dualistic thinking pushes you to refuse to settle for less.

- **Am I judging myself as strong or weak? Smart or stupid?** Remember, judging yourself is never a good idea since you'll be your own worst critic. Not being strong enough to do one thing does not mean you're weak. Not knowing something doesn't make you stupid. If you set high standards and don't completely meet them, going to the other end of the spectrum sells you way short. Since perfection is impossible, this will usually keep you on the low end of self-esteem. Good self-esteem begins with self-acceptance. You can't have that if you're always reaching for the top branch on the tree. There's plenty of fruit on the lower branches that can be just as sweet if you taste them.

- **Am I over-monitoring my decisions as right or wrong? Good or bad?** Remember, making decisions can be hard in general. When you put rigid judgments on them, it keeps you waffling about what choice to make. If you decide to do something that doesn't work out the way you hoped, it won't be the end of your world. Since there are so many circumstances beyond your control, even a good decision can disappoint you. Thinking about your potential decisions as either positive or negative keeps you from letting your gut help you. Using a dualistic mindset about making decisions can waste a lot of time and will always keep you anxious.

REPLACEMENT THOUGHTS— THE FIVE-MINUTE RULE

Once you identify a dualistic thought, give yourself at least five minutes to respond differently. Try to calm down any emotions that can influence how you judge something in black-and-white terms. When you can get into the habit of stopping to assess what your reaction should be instead of having a quick, knee-jerk reaction, you can take control of going to extremes. Change how you come to conclusions by using the following affirmations:

- **I will locate the balanced gray area of any stressful situation that's presented to me.** Remember, the gray area is there if you look for it! Bring the black and the white options closer together so you can find the gray point of view—the balanced middle ground.

- **I will be more reflective than reactive.** Remember, in almost any situation, it's advised to think before you act. The same is true about making judgments. While it's normal to let emotions stir you to an immediate response, giving yourself time to think about it allows you to find a perspective that's more balanced. That helps keep dualistic thoughts in check and allows you to find a more rational way to see what's going on and to judge yourself less harshly.

- **I will find my center.** Remember, if your mind relaxes, you get into your more balanced zone. While it might feel uncomfortable at first if you've never been there, once you allow yourself to live in the gray instead of only seeing black-or-white, you'll find it's a much more pleasant place.

- **I will accept the subtle balance and vague degrees of life.** Remember, life ebbs and flows like the tide. When you can navigate through the times that can cause anxiety, you'll feel less anxious and enjoy life more.

- **I will accept that I do not need certainty about everything right now.** Remember, what you need is to accept that everything won't always go your way and that's okay. It helps you relax about your circumstance. You can always find alternatives if you stop and look for them.

Affirming these phrases will help them to sink in. Slowly you can retrain your anxious mind into a more balanced way of thinking. As you get used to it, your thinking habits will change from black-and-white to a more relaxing shade of gray.

Chapter 5

RISING ABOVE THE
ILLUSION OF CONTROL

---

*"It's not the load that breaks you down,
it's the way you carry it."*

—Lena Horne

Most of us live with the illusion that we can somehow control critical facets of our lives in varying degrees. This kind of thinking is very seductive because it can give you a sense of security, especially if you suffer from ongoing anxiety. While this is a falsely empowering sense of security, it can make you feel safer in the world and may even help you to think you're protecting the people you love. The funny part about it is that you may really believe it works. But it doesn't.

CREATING A FALSE SENSE OF CONTROL

Thinking you have power over every circumstance in your life can motivate you to try to control everything from people to the aging process to the stock market. Sometimes, you may even try to control what you rationally know is impossible. For example, if you're driving in heavy traffic and are late for an important appointment, you may act like you can control the flow of traffic by doing things like madly yelling at other drivers, honking your car horn and flipping people the bird. Why would you act this way unless you think it will work? When the cars ahead of you have nowhere to go, do you still honk and yell? People do this even though there's no rational way that honking and yelling and getting anxious can make the traffic move.

When your initial efforts don't work, you may continue to come up with strategies to try to control something that's uncontrollable. And if you fail, and you probably will, you'll feel desperate and your anxiety will likely skyrocket. Consequently the following pattern is common:

Feeling scared can motivate you to try to control everything, but you can end up in a panic when it doesn't work. Instead, attempting this level of control is a recipe for escalating anxiety. This chapter will help you recognize this pattern and offer suggestions for letting go.

TRYING TO CATCH THE WIND

For many people, especially anxiety sufferers, the idea of letting go of control is quite frightening. You may be afraid to let your guard down because it feels like everything in your life will fall apart, or that you'll become neglectful of your responsibilities.

Whenever you think you can control something or someone and count on having that control as the sign of your success with that endeavor, the responsibility for your happiness goes outside of your reach. So, it's like trying to "catch the wind" because, obviously, the wind cannot be caught. It's impossible.

You may count on getting a specific amount of money in commissions or assume that a friend will support an endeavor that matters to you. But since these factors are actually out of your control, you'll rarely get the results you need. Your commissions may fall through or your friend may get too busy to give you the support you need. It's a formula for high anxiety because you become a victim of thinking you have control over everything in your world when you don't.

If you find yourself thinking you have the power to control everything, try to step back and be more reflective instead of reacting quickly, like I discussed in the previous chapter. Instead of striving unsuccessfully to control outside factors, you can work on the inside—making an effort to control yourself and learning to more readily accept external issues. Controlling yourself helps control your anxiety.

Before I recognized my own illusion of having control over my life, I suffered because of my strong desire to manage everyone and everything. I remember getting very anxious for hours and sometimes days before an event when I began public speaking twenty years ago. In retrospect, I was trying to control how my audience would see me. I worried constantly: *Will they like me? Will I bore them? Will they laugh at me? Will I look like a fool?* I was so fixated on the judgments that the audience might make about the presentation and me that I placed my entire self-worth as a person into their hands. Big mistake!

The truth was, even if I delivered a flawless presentation, there would probably still be people who found fault with it. Yet I turned to control strategies to do my best to be perfect and ultra-prepared on the big day. The consequences were that I didn't sleep the night before, my stomach was in knots and I lost my confidence. Why? Because I was trying to control something I had no control over by hoping to get forty complete strangers in the audience to love me, which was impossible. Once I accepted that I had no control over how people would respond, and that was okay, I was able to let go of my need to control everything. My anxiety dropped and I felt better before and during future presentations.

As mentioned previously in this book, your goal should not be to change the world or people or places—because you can't. But you can change your response to the world. The load of the world is not what breaks you down; it's thinking that you can control it and how you carry that burden. Once you let go of trying to control what you don't have control over, you can get your life back. Never forget—the only thing you can have control over is yourself. Yes, folks, that's all you can truly control! You may succeed at influencing people and work to achieve some minor successful outcomes. You may encourage people and perhaps even inspire them, but that's it. Every person is responsible for controlling his or her responses. When you modify yours, a more realistic, positive change can occur as you approach the world differently.

STRIVING TO BE PERFECT

Over the many years I've been in private practice, I've observed that when you cling to an excessive need for control as a way to try to lessen anxiety, you exert this control in ways that can appear to be manipulative and overbearing. But the two most common ways of trying to control life circumstances that I've witnessed come from an excessive need to be perfect, and relying on others for approval—or being a people pleaser.

Striving to be perfect is a dualistic concept that means having extraordinary standards for yourself and for other people that are unrealistically high. You may believe that at one hundred percent you're successful but at ninety-eight percent you're a failure. Similar to how I beat myself up for getting three out of four hits when I used to play softball.

For example, if athletes who strive to be perfect enter into competitions and win a silver medal or come in second place, they don't consider themselves a winner. Instead, they see it as losing a gold medal or missing first place. It's not uncommon to see an Olympic athlete crying over "just" winning the silver medal. As you watch, you probably think they should be happy to be the second best in the whole world because they've competed against so many incredible athletes. Yet they can't appreciate that because the pressure to be number one trumps that.

The mindset of always having to be perfect is a slavery of the mind that will never be freed. It's the ideal setup for anxiety, and if it's not recognized and cut down to size, it can become the perfect storm that will feed your angst tremendously. Making progress should be your ideal goal, not struggling to be perfect. Appreciating any progress you make feels good. The anxiety of not being perfect doesn't.

PEOPLE PLEASING

Relying on others for approval, which often leads you to become a people pleaser, also lays the groundwork for anxiety. It's impossible to have everyone love you and never be disappointed by you. Even if you lived on a tiny island, with only five other people living there with you, and you tried very patiently to make them all like you all the time, you'd probably still fail since someone will still find fault (maybe the people pleasing will annoy them!) with something about you. It's just human nature that you can't please everyone. Therefore, like with needing to be perfect, the need to be liked or loved by everyone is bound to flop. And the pressure it puts on you to keep trying to achieve it can trigger anxiety.

Being a people pleaser forces you to put your own needs and desires aside and focus on what other people want. Doing so can hurt you in many ways. If you don't get enough approval, the pressure can push you to try harder to please. For instance, let's say you have a tendency to over-extend yourself with family and friends in order to please them. Perhaps you might overspend beyond what you can realistically afford, like treating them to expensive dinners or buying pricey gifts during the holidays. Or you sacrifice much of your time for them or negate your wishes and desires to the detriment of you. Or you just might be the kind of person that says "yes" more than "no." Consequently, if your kindheartedness is not reciprocated and you don't get that validation you seek, it can leave you feeling empty inside. Hence, when you neglect yourself for the sake of others, it might feel like you are doing good, but over time, you run the risk of building resentment.

Relying on others for approval by being a people pleaser means you must be vigilant (controlling) about always saying and doing the right thing. If the need is strong, you may practically live and die by what you think other people are thinking of you. It's as if the sun rises and sets on whether or not you get validation from your friends, family, coworkers, superiors and other people important to you. And it keeps you focused on them at the expense of yourself and your own needs and desires. All of this can build large amounts of stress.

KRISTEN'S STORY

Kristen was a twenty-six-year-old woman who was living with her boyfriend, Bryan, for the last three years. They were very committed to each other and planned to marry and have children within the next few years. Kristen came to therapy wanting help with her anxiety and insomnia. She explained that she was unhappy with her job as an administrative assistant at a large construction company. It was unsatisfying and definitely not the career she envisioned when she was in school. She had studied science in college and was confused about why she didn't pursue her passion. She always wanted to be a marine biologist, yet she settled for a job that got her no closer to that goal and just paid the bills.

Kristen's anxiety had started four months earlier when Bryan admitted that although he deeply loved her, he wasn't sure if she was "the one" for him anymore. He also said that since his career was in flux, he was scared that he couldn't provide for her appropriately or be the husband he'd like to be. A few weeks later, Bryan also revealed that he had an affair with a coworker. Kristen was devastated but refused to leave him in hopes of patching things up. It weighed on her every waking moment and kept her from sleeping.

For Kristen, the uncertainty about what would become of her future with Bryan was killing her. She couldn't concentrate on work or anything else for that matter. It felt as if his ambivalence was shattering her world and making her obsessive and controlling. This resulted in her suffering from intermittent panic attacks with frightening heart palpitations and labored breathing. She sometimes secretly followed Bryan around and even checked his cell phone when he was asleep to see if he'd been speaking to the woman he had an affair with. She also interrogated him when he came home late and always suspected the worst. Yet she was terrified that she'd upset him enough to make the situation worse.

Kristen also confessed that even before these problems started, she had always been an anxious type of person who worried a lot. But once she met Bryan, most of her angst and controlling tendencies disappeared. When they first met, she described them as being blissfully connected, and they had planned almost every minute of their free time together. She tearfully asked several times in one session if I thought she should leave him now or instead try to figure him out. She was confused about what direction to go and said she was trying harder and harder to please Bryan but didn't know if it would be enough to keep him happy.

Kristen discovered that the source of her severe anxiety came from her excessive need for control. She recognized that by trying to control Bryan and his erratic behavior, she was making herself feel worse.

She begrudgingly acknowledged that despite the betrayal, she was trying to fix the relationship by using manipulation and spy tactics as a desperate attempt to save it.

In order to cope with life's road bumps, Kristen began a pattern of trying to control the people around her and her environment, believing it would make her feel safe. Because she was so fixated on everyone else, she seldom lived in the moment so she could enjoy her own life. After she got involved with Bryan, she became obsessed with wanting to be perfect for him and to please him all the time. This obsession was actually an illusion of having control. Kristen fell victim to this false belief by adhering to a very specific inner dialogue that she was convinced would allow her to feel safe by controlling her life:

- *To feel safe in this world, I must have complete certainty about everything.*
- *To feel safe in this world, I should micro-manage everything and always be vigilant.*
- *To feel safe in this world, I should pay close attention to everyone's behavior.*
- *If I am rejected by anyone, it will mean that I'm a worthless person.*
- *If Bryan leaves me, nobody will ever love me again and I'll always be alone.*

She repeated her inner dialogue as if it were a certainty. Notice how she used the kind of absolute words discussed in Chapter 3, such as *should, must* and *always*.

Over time, Kristin let go of these beliefs. She let go of being the fixer and the people pleaser and began focusing on herself instead of trying to control what was outside of herself. She learned that *her* needs had to come first in order to be happy and she *did* have control over that. Kristen also accepted that she had zero control over Bryan or anyone for that matter. I helped her to see that the strategies she chose to help her feel safe only made her feel worse. So by giving up the need for control, she subsequently stopped feeling out of control and got her life back on track in a more satisfying way.

ACCEPTING THE LIMITATIONS OF YOUR CONTROL

I will use our solar system as an example once again to give you another picture of your place in the world in relation to other people. Hundreds of years ago, the accepted truth about the configuration of the planets was that the Earth was the center of the universe and responsible for the gravitational pull of all the other planets. The Earth was considered the centerpiece that everything revolved around. People believed that if for some reason the Earth shifted from its position as the heartbeat of the solar system, the planets could dangerously realign and go off course from their assigned orbits and there would be complete interstellar chaos. Today we see things quite differently. What astronomers now believe is that Earth is actually just a small planet that orbits around a sun that's vastly greater than it. And the Earth isn't responsible for much at all, except maybe for the minor duty of being the gravitational pull of our tiny moon. The world doesn't revolve around you either.

It's important to try to remember that you're not the center of anyone or anything's universe and you are just one diminutive planet out of many. The reality is that you only have your own orbit to worry about. When you accept that, you can let go of the assumption that you possess any kind of power or control over anyone or anything. You simply don't have that kind of supremacy. You are *not* the sun!

Remember, you'll be far less anxious if you let go of the illusion that you can control everything and focus on altering your *responses* to life, instead of failing miserably at "trying to catch the wind." Over time, you'll begin to see that, like Kristen, you indeed have the power to change your own reality by focusing on you, not on other people. It begins by accepting that the only thing you have control over is you and how you react to what's going on in your life. And that's fine.

PROCESS ORIENTATION VS. RESULTS ORIENTATION

Another helpful way to get beyond the illusion of control is to focus on a *process orientation* way of thinking—accepting that life is a process that will change and bring unexpected situations, and finding ways to deal with what you're hit with—instead of just focusing on the *results* you want in a situation, which I refer to as a *results orientation*. Anxiety sufferers that I've treated are astonished when they realize how much angst they've caused themselves by adhering to the latter. Results orientation is a mindset based in the fantasy that you must have control, and you continually seek guarantees that everything in your life will work out as you want it to. So you always focus on the results that you're seeking. A results orientation to living means you:

- Worry a great deal about the future.
- Believe that success is the only way to measure value or worth as a person.
- Want to know that you'll never fail.

- Beat yourself up about things in the past.
- Are a perfectionist about yourself and others.
- Have unrealistic expectations about things (like feeling that you only succeed at 100% but fail at 98%).
- Want to know that you won't have to feel any pain.
- Believe that you need to please everyone all the time.
- Stay attached to the irrational belief that you can get guarantees for everything in life.
- Want to know that all the people you love will always be safe.

A consequence of this results orientation mindset is that you might become a people pleaser or codependent to your family and to other people in your life—someone who focuses on other people's needs to the detriment of yourself in order to be loved. And, in the arena of your career, you may become an overachiever or a compulsive personality with very poor impulse control. Results orientation will turn you into a "human doing" instead of a "human being."

American Buddhist and author Pema Chodren said, "Everything is in process. Everything—every tree, every blade of grass, all animals, insects, human beings, buildings, the animate and inanimate—is always changing moment to moment. It means life is not always going to go our way. It means there is loss as well as gain."

Hence, when you cultivate and maintain a process orientation, you're engaged in life but in quite a different fashion than when you always try to get the results you want. You may be plugged in to the same ideals and goals, but with process orientation, you are more rational and patient, and you understand that nothing is permanent and that everything in life may change all the time, which is a normal part of life.

Process orientation means that you learn to accept the uncertainties of life and grasp the often-frightening reality that there are no guarantees for anything. That allows you to actually seek comfort in the fact that everything is forever changing so you can give up searching for comfort in the illusion of permanence. When you have a process orientation mindset you:

- Don't try to overmanage outcomes.
- Live in the moment and don't focus on thinking about the future.
- Don't dwell in the past or try to change or improve it.
- Focus on what you can do to have control over the present.
- Understand that there are steps required to achieve goals and that it's up to you to take them.
- Don't use manipulation to go after approval from others.
- Develop realistic standards for what success is for you.
- Acquire the tools to discover the gray areas of life.
- Accept the uncertainties of your life.

As a psychotherapist in training many years ago, I remember getting easily attached to results orientation by wanting to "fix" every patient that I treated. It was a noble cause of course, but I found out that thinking this way caused me a lot of anxiety because it was unrealistic and impossible. I was very focused on the outcome and I wanted to make people happy quickly without appreciating the process. I wanted to soothe their pain instantly and cure them of their maladies. But I slowly learned, as all therapists do, that the process itself is the most valuable part of the therapy and not a quick solution, because there is no quick solution.

Without this insight, I would have become a frustrated clinician and burned out a long time ago. I would not have survived putting that kind of pressure on myself. So even as a therapist, it's impossible to be a fixer, but it's a little easier to be a facilitator. I learned that I was not responsible *for* my patients (because I cannot fix them) but I was responsible *to* them. I was responsible to them by empowering each and every person that walked through my door to build solid coping skills and to make better decisions in their lives. To use a metaphor, I finally got comfortable teaching my patients how to fish without catching the fish for them.

Therefore, process orientation also means that you are a facilitator in the progress of your life, not a fixer. When you act as a coordinator for the way your life unfolds, your efforts go to creating the conditions needed for success and happiness to blossom. By doing this, you put aside the seductive quality of the outcome in favor of grounding yourself in the here and now, which is the only gray area that you have some control over. And you don't react when things go wrong by reaching for empty fixes.

A wonderful example is Lance, a patient I used to treat for severe panic attacks. He was tangled in a nasty web of results orientation. On the first day of his treatment, he said to me frantically, "If I could just stop worrying about money for one day, I would feel so much better." Lance's statement illustrates what can happen to someone who is very obviously concerned about making and saving money. He reached the point where the results orientation mindset that kept him trapped in his money beliefs also tricked him into thinking that he could make his anxiety go away and his life would miraculously improve if he could have one day of peace.

Lance focused on getting that one special day of not having to worry about money and saw it as his oasis and the answer to alleviating his anxiety. Yet he could never get that one worry-free day. His attachment to his outcome was so fixed that he irrationally believed it could be eradicated as simply and quickly as getting that day. He was so desperate to feel better that he reached for the only thing he knew, which was the fantasy of a swift and unrealistic fix. He had to work to let go of that belief and learn how to use the techniques in this chapter to find a more realistic goal.

DWELLING IN THE PAST

Process orientation also means that you're not particularly concerned with dwelling in the past and trying to uncover the origin of a problem. Even though Jean-Paul Sartre said, "all introspection is retrospection," and looking in on the past can indeed help you to understand yourself and heal old wounds, sometimes this search can cause you a great deal of distress. Author Robert D. Zettle's "flat tire" analogy helps to illustrate this wonderfully.

Zettle asks: if you're driving in your car and suddenly get a flat tire, what's the first thing you'd do after you pull over? Typically most of us will do whatever we can to remedy the problem and get on our way. You may pull out your jack and begin changing the tire yourself, or perhaps more realistically, take out your cell phone and call for roadside assistance. You probably would *not* immediately go back along the street or highway you were just on to look for what caused your flat. You wouldn't spend an extravagant amount of time scouring your previous road looking for the nail or the shard of glass that popped your tire. Why? Because in the moment, you won't really care.

You'd be more focused on your day and where you're going. Even after you take the flat tire to a service station to get it fixed, you probably won't be too concerned about the how and why the tire failed and more likely would just buy a new tire. Letting go of the illusion of control is sometimes also about letting go of the "why" and the "how" and not dwelling in the past for answers to everything. In fact, sometimes the past has no answers, and getting too caught up in it can slow you down and make you feel more anxious. Wanting these kinds of answers goes along with a results orientation, which, as discussed, can leave you feeling powerless.

TOOLS FOR LETTING GO OF CONTROL

It can take time to let go of your need to believe that you can control the people and circumstances around you, but it's worth the effort. It begins with raising your consciousness about your patterns. Sometimes "letting go" is the best way to hold on.

Letting go of your need to control everything can be one of the scariest habits you can break. You may worry that you'll be judged for not being perfect or that you'll alienate the people you want to like you. But most find that life is just fine once they make the change because having less anxiety feels so good. Even if the idea of letting go is uncomfortable, it's important for you to release that need to control other people and circumstances and instead take control of *yourself* if you want to reduce your anxiety. The more you recognize the behavior discussed in this chapter, the more power you have to control your future responses. To identify areas that you try to have control over, ask yourself:

- **Am I trying to be a fixer in my life?** If you think that if you try hard enough you'll be able to fix everything in your life that's not perfect, you set yourself up to be disappointed, frustrated and angry. Anxiety can increase as you try harder and harder to change what's not working. You need to recognize that you can't fix your life by controlling everything, and that's okay.

- **Am I preoccupied with outcomes and excessively worrying about the future?** Since the future isn't here yet, you can't know what will happen. Worrying about it by thinking of what you can do for any worst-case scenarios you come up with just enhances anxiety. When you can retrain your mind to focus on the present, you take control of a big anxiety trigger.

- **Am I expecting things to happen too quickly?** Patience is called a virtue for a good reason. Wanting instant gratification all the time won't bring you what you want any faster than patiently waiting for the right time or circumstances for it. If you don't get things right away your anxiety may escalate with the thoughts created by the emotions you get when you want something to happen and it doesn't appear in the time frame in which you want it.

- **Am I trying to control other people's feelings and actions?** It's important to recognize this pattern so you can stop it because the only person you can control is yourself. Trying to control other people leads to frustration when they don't do what you want, which creates anxiety. When you control your own response to people's behavior, you can often get more satisfying results.

- **Am I afraid that if I let go just a little bit I will lose everything?** Fear can make you want to keep every little thing under tight control. You need to keep all your ducks

in a row, so to speak, in order to keep an eye on everything
and make sure it all goes the way you think it should. But
since most things aren't in your control, not letting go at
all will wind you up tight and keep your anxious mind
working overtime.

- **Am I seeking approval from others in order to feel good
  about myself?** If you can only see yourself through the eyes
  of other people, you won't feel good about yourself inside.
  It will keep you tense, ready to jump to do favors and other
  things to keep people around you happy in order to get
  their approval. You'll always be at least somewhat on
  edge, worrying if you're doing enough. People pleasers
  are rarely happy, except for the moments when they get
  thanks or some kindness from someone they pleased.
  But it's usually fleeting.

REPLACEMENT THOUGHTS—
THE FIVE-MINUTE RULE

As I said, letting go of your need to feel in control of everything can
take time because it can feel scary and leave you feeling vulnerable at
first. But once you do it, you'll see that the rewards will give you more
real control over yourself. Take five minutes to respond differently
after you identify a controlling thought or controlling behavior. As
they come up, use the following affirmations:

- **I will allow life to unfold without me trying to control it.**
  You can do this! When you start to go with the flow instead

of trying to swim against the tide, you'll be able to relax more and let life happen instead of trying to make it happen the way you think it should.

- **I will accept that being a facilitator instead of a fixer will reduce my anxiety.** Trying to fix everything increases anxiety. Working with the reality of your life and controlling your responses to it allows you to make the most of what you have or the circumstances around you, which can lead to increased happiness.

- **I will accept that worrying about the future only makes me more anxious.** Worrying about what hasn't happened—fear of the unknown—increases overall anxiety because it's projecting into the future, and there will always be a future to worry about. And the unknown you worry about is often much worse than what actually happens. When you focus on now, and know you're fine, you'll feel a lot more relaxed. You can deal with tomorrow when it's here.

- **I will accept that the only thing I have control over is myself.** You can keep fighting that fact or accept it. Trying to control situations or people can drive you crazy when they refuse to adhere to what you expect or when things just don't go your way. Focusing on yourself actually gives you a lot of control because responding to situations and people differently can get better results than directly trying to control them.

- **I will accept that "letting go" does not mean giving up or that I will lose things.** Life will go on. People will have their own opinions. But you don't have to feel the pressure to change them. The main thing you can lose is a lot of the anxiety you've been feeling.

- **I will focus on process orientation instead of results.** Learn to work with what you have! Practice accepting life the way it is, not how it should be, and remember that there are no guarantees about anything. When your expectations become more realistic and your patience increases, you can live in a less anxious mindset.

- **I will accept that I don't need approval from others to feel good about myself.** You really don't. Start looking for your good qualities. You have them. In Chapter 8, I present tips for building your self-esteem. For now, just accept that you can learn to approve of yourself so you don't have to buy approval with favors and people pleasing.

Chapter 6

THE ACCOUNTABLE SELF

Taking Responsibility for You

*"Man is nothing other than his own project.
He exists only to the extent that
he realizes himself."*

—Jean-Paul Sartre

According to existentialist thinkers like Jean-Paul Sartre and Søren Kierkegaard, we're born into a cold and empty universe with no meaning. When humans fully comprehend this void and isolation, the "Self" is born. The practice of living existentially is a process of defining the essence of who you really are and accepting your unique place in the world.

If you can strip yourself of the belief systems you were given, you can cultivate your own personal sense of self. Then you can choose to freely decide how you want to live and who you want to be, which is the essence of living existentially. According to Dr. Irvin Yalom, "Each of us must decide how to live as fully, happily, ethically, and meaningfully as possible." When you let go of the beliefs that don't serve you well, you can be the navigator of your life instead of living on autopilot with your world dominated by beliefs from others.

LIFE AS A POKER GAME

Life can become like one big poker game when you choose to make it yours. There are many different ways to play poker, but generally you begin to play when you're dealt a particular hand of cards. The draw of these cards is random of course, so you have no control over what's handed to you. But ultimately as the game progresses, you must choose to do something with the cards after you look them over and consider what you have.

For example, you could be born with a "good health" card or with a "chronic illness" one. You could be born into a life of privilege or a life of poverty or somewhere in between. You could be born into a family with abusive parents or into a kind and loving family. What your sense of self does with these life cards is totally up to you.

There are other people playing in the game, too. You must consider countless possibilities and consequences based partly on your life cards and also on your in-game decisions in relation to the other players. Without the other players, there is no game. But remember, to succeed at poker or to travel through life relatively unscathed, you can't just rely on getting lucky life cards. You may get a few good hands dealt to you here and there, but it probably won't last. There are no guarantees. The skill you acquire in playing your life cards wisely is the key to survival.

I didn't always play my cards right when I was young. I tried to avoid my anxiety card at all costs by making an effort to hide it and pretend it wasn't there. I was so ashamed of that card that I didn't even tell people I had it. Finally, I decided that instead of running from it, I took responsibility and played the card by getting therapy for it. Later, I even played the card to *my* advantage and decided to use my anxiety as a springboard to a career as a therapist. If I had not used that card to become a therapist, I still might be feeling sorry for myself for having been dealt such an undignified card. I took responsibility for *me*, despite a chronic illness card, and made lemonade out of lemons.

ALLISON'S STORY

Allison was a twenty-eight-year-old single female who came to therapy seeking help for anxiety and insomnia. For the last five years she worked in sales for a pharmaceutical company. She didn't particularly enjoy her work and knew it wasn't her calling. The feelings she had about her job ranged from boredom to being totally uninspired, but she stayed because she needed to be able to pay her bills. Allison admitted that she had always wanted to be a doctor but didn't think she was smart enough to make it through medical school. Years before a friend had told her that she didn't have what it takes to be a doctor. That statement stuck in her mind and greatly influenced her decision to not pursue a medical career.

Allison also admitted that, for most of her life, she looked to people for guidance and was often dependent on others for validation. Unfortunately, they weren't supportive of her desires and never encouraged her to go further in her career choices. What they thought about her strongly influenced her decisions, especially discouraging her from trying to become a doctor. She believed that certain things in life were just meant to be and that this job, as much as she disliked it, was as far as she'd ever go in her career. Allison gritted her teeth and became determined to stick it out so she could support herself.

In our second session, Allison said that her mother, whom she was always very close to, had recently died after suffering a stroke back in Wisconsin, where she'd grown up. She didn't make it there in time to see her before she passed away. Although her mother was in a coma after her stroke, Allison described feeling immense guilt about failing to return quickly enough to see her before she died. She'd felt obligated to deal with a crisis at work, which delayed her from going. Allison also felt very guilty that she had chosen to leave Wisconsin seven years earlier to start a life in Los Angeles. She experienced a great sense of remorse about her choice to move and leave her mother and strongly believed that, as the only child in the family, it was a very selfish move.

Allison actually convinced herself that if she had stayed, maybe her mother would still be alive. She also shared that for the last two years she was in a verbally abusive relationship with a man whom she was afraid to leave. Her dependency on him made her feel trapped and sure that she couldn't survive without him. Allison admitted that she was very timid and submissive around her boyfriend. She neglected her needs and deferred to him most of the time. Her eyes teared up as she expressed terrible shame about her choice to stay with this abusive man and how very weak she was for letting it drag out for so long. She knew she should leave but felt stuck.

Allison eventually recognized that she was naively convinced that her current state of mind and the situations she was in were foregone conclusions and set in stone. This convinced her that it wasn't possible to create her own identity and pursue becoming a doctor. Her belief that certain things in her life were or weren't meant to be reflected that she wasn't taking responsibility for the quality of her life or for the choices she made that were necessary for her to mature as a human being. Her decision to leave Wisconsin years ago was made because there were no decent-paying jobs in the small town she was from. She chose to leave in order to better herself.

Her intentions were good and she was trying to be responsible for herself. But when she arrived in Los Angeles, she had let others steer her course. Allison also recognized that the anxiety she suffered was rooted in her inability to be an authentic person in her interpersonal relationships with people, especially with her abusive boyfriend. She allowed herself to be victimized by him, which kept her anxious mind working overtime. Another source of her anxiety was her inability to accept the choices she made seven years ago when she decided to move to Los Angeles and, most important, the choices she was too afraid to make for her career going forward.

Allison suffered from what I label as an existential crisis because she couldn't reconcile decisions made in her immediate past and was immobilized by her inability to take even a baby step to creating a better life for her future. She was too frightened to play the life cards she was dealt. As a result, she became a victim of her external world, which left her feeling helpless to think positively and take action that would improve her life. Allison was not being accountable to her true self because of the inner dialogue she lived with that kept her mind anxious:

- *I am weak and I can't change.*
- *My life is already laid out for me and there's no reason to dream of doing more.*
- *I am helpless in relationships, especially intimate ones.*
- *I am selfish for thinking of myself.*
- *I am a bad daughter for choosing to leave Wisconsin.*
- *I am partly to blame for my mother's death.*

Allison's crisis began to disappear when she recognized that to be happy she would have to play some of her cards and risk the possibility of failure and perhaps pursue a medical career despite what her friend had irresponsibly told her. About a year later, she enrolled in medical school to pursue her dream, and her life changed dramatically. It changed because now *she* was holding herself accountable for her happiness. If it failed, she could at least say that she tried and would have no regrets. Allison was much happier with her life, and soon after beginning classes, she broke up with the abusive boyfriend and got her own apartment.

DEVELOPING AN ACCOUNTABLE SELF

It's important to recognize the importance of being accountable for how you handle the cards you've been dealt. When you have an accountable self, you play the cards the way you want instead of letting someone else dictate what to do with them. For years, I deferred to others to make important choices for me—especially my father—because I was too afraid to hold myself accountable for the consequences of those choices. I let people make decisions for me that, at times, were very unwise decisions. I grew up in a home where I was taught to doubt my uniqueness as an individual and to doubt my place amongst my peers. However, once I discovered that deferring was causing me more anxiety, I stopped relying on others and took more responsibility.

Taking responsibility empowered me to trust myself and to feel like less of a victim. Existential elements force your accountable self to develop due to some of the following qualities that offer many beneficial results:

- **Search for personal meaning:** The search for personal meaning is never ending. Being accountable means you're constantly asking, experiencing and transforming yourself as you adapt to life's circumstances and relate to others. Life is not meaningful enough on its own—you must give it meaning! In this ongoing search, you can learn, grow and mature as you create your own reality. You don't have to adhere to fixed beliefs about yourself, others and the world around you. You can engage in process orientation and exercise your ability to make choices in your life that are based on a value system that you, not other people, cultivate.

  Being human means discovering and making sense of your existence on a regular basis. The enlightenment that sustains you as a healthy person lies in developing self-awareness and the search for answers and choices that truly feel right to you, not just to someone else. It's up to you to find a real personal meaning based on your needs and what makes you happy, not what other people try to tell you to believe or that you "should" adhere to.

- **Self-awareness:** When you choose to become accountable, you'll need to develop your self-awareness and a desire to examine your life. Socrates said, "The unexamined life is not worth living." That said, to have a worthwhile life, you should never stop questioning and challenging all the previous meanings and belief systems you've been raised with. It's time to start to formulate your own. But first you must also know yourself and understand and accept your limitations as a human. You can learn to adapt and struggle with forces beyond your comprehension. And it's important to understand that you cannot and will not ever know everything, and that's okay.

  You may also experience loneliness, alienation, guilt, fear and anxiety. To take responsibility means to accept that and act accordingly. That means you let go of trying to control other things or other people and accept that sometimes experiencing uncomfortable emotions is part of being human. You can try to understand yourself in the best way you can, with compassion and respect. Descartes proposed the challenge, "Conquer yourself rather than your world." You can be your own worst enemy or your own best friend. The latter will help reduce your anxiety.

- **Personal responsibility:** Taking personal responsibility for the choices you make also motivates the formation of your authentic sense of self because you stop blaming others and stop playing victim to circumstances. As you do this, your inner self learns to define itself based on its own sense of what it needs. It doesn't rely on others to tell you what to do. So you'll hopefully reach the place where you're compelled to make authentic and independent choices because no one will do it for you. If you do let others take charge of your direction and how you respond to life's situations, you won't be living with an existential perspective and most important, you're not living as your true self.

Sartre said, "You are free to choose; in other words, invent. No general code of ethics can tell you what to do. There are no signs in this world." And even if there are signs in your life, the truth is that you choose how to interpret those signs. Taking responsibility for the circumstances in your life and holding yourself accountable at all times might feel frightening, especially at first. But if you think about it, it's actually more liberating to know that you have some power over what may make you suffer, especially your anxiety. Without the power to take responsibility, you feel powerless. Taking responsibility gives you the power to control more of what you don't like in your life so you can improve it.

- **Meaningful relationships:** We're social creatures who like to have relationships with other people. But identifying your authentic self within interpersonal relationships is not easy to do. While you can't escape the interdependence you'll always have with other human beings, you can forge a solid sense of self within that interdependence. If you do not forge that sense of self, you may become the sum of other people's expectations of you or a victim of negative beliefs that your parents or other influences ingrained in you. They may even exist only in your subconscious, but they still guide your actions.

You could easily become a people pleaser and alter your personality at any moment to fit it in to what your family and society want for you. But you can also push for connectedness and intimacy while still maintaining the core issues of your true self. It's a choice that you as an accountable person must make. How much will you choose to absorb from others or to deflect and own yourself? Your sense of self acknowledges and respects the individualistic spaces that separate us as humans. It also understands where the "I" ends and "you" begins. Meaning the "I" is not an extension of someone else. It is independent. Those boundaries can take time to establish, but you can have them with lots of awareness and a desire to be accountable for your life.

A good example is Beth, a thirty-nine-year-old wife and mother of two, who came to me complaining of panic attacks. She wasn't sure why she had them, but she believed they had to do with her marriage. After a few sessions, I recognized that Beth struggled to establish a solid sense of self with her spouse because she was afraid that if she asserted her individuality, he'd leave her. Beth learned at a young age from her dominating and chauvinistic father that all women must be subservient to their husbands. She essentially had zero personal boundaries and didn't take responsibility for herself in the relationship with her husband. Consequently, she was very unhappy and admitted that the marital discord between them got worse in the last few months when her husband became manipulative with her and also verbally abusive.

Because Beth was a bit of a people pleaser with no sense of self in the marriage, she had no idea how to deal with it. That left her rattled with anxiety, and it scared her that her marriage was falling apart. In our work together, she finally understood that she had no voice in the relationship—no way to express her thoughts, feelings and wishes to assert her individuality. She wasn't taking responsibility for her place in the family as a wife and mother. Without this personal accountability, she was just another one of her husband's children or still like her father's child instead of being his adult daughter with her own individual identity. Once she realized this, she began to modify her position in the marriage and, over time, her relationship with her spouse drastically changed and, most important, her anxiety decreased.

ACCEPTING DEATH AS INEVITABLE

Eckhart Tolle said, "What will be left of all the fearing and wanting associated with your problematic life situation that every day takes up most of your attention? A dash, one or two inches long, between the date of birth and date of death on your gravestone."

I'm sure that you're aware that you'll die one day, just like everyone else. For most, it's a scary and disheartening fact of life, one that people typically choose not to think about very often. But if your inner self is prepared to view death as a condition of living that inspires further awareness and positively augments the beauty of life, then death is merely a tool—a value-enhancing step in life's journey—not something to be afraid of or dreaded. Awareness of knowing there will be an end to your life gives your life meaning. It thrusts you into giving the time you have on this Earth significance and purpose.

Knowing your end will come one day helps you to better appreciate your relationships, your career and all the other good things you have in your life. Lastly, everything that you love and cherish, everything that you value with tremendous passion and gravity, is anchored only by the sobering awareness that you can lose it all very easily and quickly. Without this mortal fragility, the things you value would, in fact, not be valuable and life would not be worth living. That's why it's so important to value your life enough to allow your own choices to dominate it so you can give it the best chance to be happy, with less anxiety.

TOOLS FOR BECOMING AN ACCOUNTABLE SELF

In order to become accountable to yourself, you first need to understand what that means in relation to who you are now and identify all the areas in your life where you're not accountable—where your mindset is someone else's. That helps you see how your desires compare to how you make your choices now. If necessary, write down your true preferences in list form. Then think about what beliefs or which people you need approval from that currently keep you from making those your realistic goals.

Being an accountable person means you own your decisions and how people treat you. You also know that only you can change what you don't like. Pay attention to the times you feel guilty or blame others for what goes wrong or any other negative emotions that come up regularly. Once you see the ways that others control you, and what your true self would like, you can take steps to claim the power over your life. In order to do this, ask yourself:

- **Have I decided that I can't ever change or recreate myself?** You can get so used to the one way you think and to automatically following the beliefs you've had for years that you convince yourself it's the only way you can be. But it's not! It's always a choice. It can be hard to let go of the thoughts and behaviors you have on autopilot, but if you look at your list of how you'd like your life to be, you can motivate yourself to at least consider testing the waters of thinking for yourself.

- **Am I just an unconscious, walking impulse, lacking awareness of my thoughts, feelings and actions in life?** Habits can do that to you, especially when they're reinforced by a need for security. That's why it's so important to pay attention to your thoughts so that you become conscious of them. Don't get angry at yourself when you identify beliefs that don't serve you well. It's not your fault. You didn't know better. Now you do, and now you can change the dynamics of how you think and live.

- **Am I playing the victim of my life circumstances and blaming others for my problems in the past?** When you're not accountable, it's easy to fall into a mindset of feeling like a victim since you believe that other people are steering you. But you must accept that you don't have to let them if you want to be in more control of your life by taking responsibility for it. Feeling like a victim is stressful. Make a note of every time you feel like one or blame your life on other people. Your anxiety can decrease once you let go of feeling like a victim because you'll understand your power to change.

- **Am I not taking responsibility for the choices I've made and for my life's problems today?** Ask this out loud if necessary. Until now, you may have assumed that you weren't responsible. This can be especially true if you equate being responsible with pleasing others and doing what they expect of you. You may have convinced yourself that this is what you're supposed to do, even if it makes you anxious. You can shift this belief and begin to be truly accountable to your authentic self. That can eventually help your anxiety settle down.

- **Am I too afraid to be authentic and true to myself in relationships, including my intimate ones?** Often when you have someone in your life who you believe you need, or if you're in love with someone, you feel pressure to please the person. You might stifle your opinions, your feelings about what annoys you or your preferred behaviors if you're concerned it might turn the person off. You may hesitate to speak your mind or express your needs for fear of losing the person. So you live with a façade of being one way while your true self stays dormant. People in love tolerate a lot of unacceptable behavior to keep a relationship going. But not being true to yourself in a relationship, whether it's with a friend, family member or romantic partner, can cause great amounts of stress. It's better to get comfortable with who you really are and accept that people worth keeping will accept you. That helps to calm anxiety.

- **Am I too afraid to accept the fact that my life is finite and that I will inevitably die?** Death is inevitable. It can be a sad and sobering thought, but it also can motivate you to live your life to its fullest. Worrying about dying doesn't help you feel good and increases your anxiety. When you focus on taking charge of your life so you can enjoy your time instead of dwelling on future outcomes, stress goes down and happiness can rise.

REPLACEMENT THOUGHTS— THE FIVE-MINUTE RULE

When you decide to become more accountable by taking responsibility for your life, you'll become more conscious of taking some breaths before you respond to whatever is going on. This way, you can catch yourself resorting to old patterns of blame, guilt and pleasing others, and remind yourself that you want to make a conscious choice to take more responsibility for whatever is happening around you. Accountability can feel less stressful than always trying to live up to other people's expectations and standards. Take five minutes to respond differently after you identify a thought that's not in alignment with being accountable to yourself. Get into the habit of using the following affirmations:

- **I will accept that I am forever changing and creating myself and that I will never be a fixed or foregone conclusion.** Allow yourself to get excited about being true to yourself, and what you want to believe or do. This allows you to have new experiences and break free of patterns you've been stuck in. Having new possibilities feels so much better than feeling stuck in ways that don't serve you well.

- **I will use mindfulness and active awareness skills to make conscious meaning out of life's experiences. I have control over my thoughts, feelings and actions.** Consciousness is a powerful tool for making change. Keep reminding yourself to stay aware of what you think, feel and do so you can identify patterns that need to be altered.

- **I will take full responsibility for all the circumstances in my life by doing what I can to make them better today. There is no one to blame anymore.** When you make a conscious decision to take responsibility for your life, and affirm it often, you can make your life better. Put your energy into improving what you don't like instead of looking for someone to blame. Accept that not everything will go your way, and that's okay. Take blame out of the picture, and don't transfer the blame onto yourself. People sometimes do that when they realize they let people orchestrate their lives. You didn't know better. Just do your best and allow your anxiety to ease instead of looking for someone to blame. But always remember that life is unpredictable most of the time.

- **I will be more aware of the choices I make today and accept that I alone am the author of my destiny.** The more you pay attention to your choices, the more they will be in your consciousness. The more they're in your consciousness, the more likely you'll be to start evaluating your choices based on what's best for you, not other people. Those choices can lead to a more authentic self and less anxiety.

- **I will remember who I am in relation to others and be an authentic, separate individual. I will differentiate myself**

from others and let them experience the uniqueness of who I am. Give yourself permission to be your own person instead of trying to dance to everybody else's tunes. You can fit in with others while being true to you. Let who you really are, or who you'd like to be, shine through for people to see. Enjoy it! Some people may need time to get used to your new persona, but they will over time. It can lead to you finding the kind of happiness you didn't think was possible for someone like you. And happiness can keep anxiety down.

- **I will use the rest of my time in this life wisely and accept that the prospect of death is a concept that actually helps me live life more fully.** Make the most of the life you have. You have a choice—enjoy it or not. You can either worry yourself into a state of anxiety that makes you suffer, or you can be true to yourself. Yes, you make the choice. Which sounds better to you?

Chapter 7

CREATING YOUR OWN REALITY

Practicing the Inner Management of Yourself

*"Your happiness ultimately arises
not from the circumstances of your life,
but from the conditioning of your mind."*

—Eckhart Tolle

As you begin to heighten your awareness about the presence of any fixed, negative beliefs about yourself that you may have, you can now turn to the actual practice of managing your inner self. These fixed, negative beliefs include those that originate from the rigid mindset I've discussed in previous chapters: the perils of clinging to consensus reality, the limiting vision of having a dualistic mind, the deceptive need for control and the hazards of not establishing yourself as accountable. As you practice managing your inner self, your reality can change for the better. As it does, you should also be able to manage your anxious mind in better ways.

IDENTIFYING ANXIETY-PRODUCING THOUGHTS

Creating your own reality starts with the delicate practice of identifying your automatic thoughts that work in conjunction with the four mindsets mentioned previously. Once you're able to accurately identify and label these thoughts as negative beliefs that you have assumed or created for yourself over the years, they can begin to feel less true. Acknowledging that you don't need or want them can accelerate the process. At that point they'll gradually weaken in strength and potency. As that happens, you'll start to see them for what they really are—beliefs that don't serve your best interests and that create anxiety—instead of blindly accepting them as concrete and factual.

A key to creating your own reality by managing your inner self is learning how to reframe the kinds of thoughts I've discussed. It can take some time and practice to create a new habit about how you perceive life's situations. But the more you practice doing it, the more you'll get used to it. And when you experience the benefits of doing so, you'll be motivated to continue to practice the new habits. The following sections outline some of the common anxiety-creating negative thoughts. You may not have these exact thoughts, but these examples should be able to give you ideas for what yours are.

This chapter will help you to identify the thoughts that can hurt you and then illustrate how to reframe or replace them in ways that reduce anxiety. Write down any of the thoughts you see in these sections that are similar to yours and any other ones that come up as you read through them. Those are the beliefs you can change to create your own reality.

EXAMPLES OF THOUGHTS THAT CLING TO A CONSENSUS REALITY AND CAUSE ANXIETY

- **There is only one way to do things so I must stick to what I've been taught to believe.** Even if my beliefs don't feel good and holding fast to them increases my anxiety, I must follow them.

- **I should have a purpose in life.** And it should be one that's acceptable to my family, even if it's not what I'd ideally like. That's what responsible people do. If I don't have a purpose, then people won't take me seriously.

- **I should be more productive/creative/ambitious.** This is what a conscientious person does. I have to prove that I can step up and get things done. Being mediocre means I am a loser.

- **I should follow in my father's or mother's footsteps.** I know they want that so it would please them, even if their footsteps don't fit the ones I'd like to take. I'm sure they wouldn't approve of the career I'd love to pursue, so I probably won't try it. I don't want to disappoint my family.

- **I should be married with children.** Isn't that what grown-ups do? I haven't met the right partner and feel a lot of pressure to find him or her. I'm not even sure that I'm ready to have children yet or to even be married. But I know it's what I'm supposed to do.

- **I'm supposed to be happy.** So I will put on a happy face, even though I'm not feeling it. I feel anxious, not content. But I hide it well and smile to cover my pain. I don't want people to worry about me or to think I am a downer.

EXAMPLES OF THOUGHTS THAT LIMIT YOUR THINKING VIA THE DUALISTIC MINDSET AND CAUSE ANXIETY

- **If I am not making X amount of money every year, I'm a loser.** I've been working hard, yet I'm not making the kind of money that my friends and siblings are. I get embarrassed when they compare salaries or when I pass on doing something because I can't afford it. Even though I get high praise at work, I know I'm no good because my income is lower than it should be.

- **If I ask for help and go to therapy, it means I am a weak person.** Isn't therapy for crazy people? If I go it means I can't handle my own problems and I should be able to. It means there's something wrong with me. I already feel wrong and don't need to feel more damaged.

- **If I make a wrong decision about anything I do, it means I am stupid.** It doesn't matter how well I do most of the time. I should get it right every time. Stupid people get it wrong. My father seems to always get it right, or at least he says he does. If I were smart, I'd always know the best choice. So I'm basically stupid.

- **If I don't go to the gym every day, I am lazy.** I can gain weight easily and know I must exercise every day or I may get fat. Sometimes I'm exhausted after work but hate the idea of seeming lazy so I drag myself to the gym. Even if I don't feel well, I won't allow myself to skip it. Since I meet up with friends there, I'd be ashamed to skip a day. I refuse to be lazy!

- **If I make any kind of mistake or forget something, I am irresponsible.** I was brought up to believe I should do everything to perfection. Mistakes mean I'm flawed. Making mistakes means I didn't try hard enough and was negligent about what I was doing. I'm ashamed of not getting it right every time.

- **Getting a divorce means I have failed.** I should have tried harder to keep my marriage intact. I should have found a way to make my partner happy or to be happier with him or her. Marriage is supposed to be forever. Something must be wrong with me if I couldn't make it work. I'm a failure.

EXAMPLES OF THOUGHTS THAT FALL VICTIM TO THE ILLUSION OF NEEDING CONTROL AND CAUSE ANXIETY

- **I have a responsibility to fix other people's problems.** That's what we're supposed to do. How can I ignore someone who needs me? I try to fix people I date, friends and family members. I even pick up the slack for people at work without taking credit. I'm supposed to know what to do. That's how my parents were and that's how I must be. I end up neglecting myself, but that's just the way it has to be.

- **I must be sure that everyone I love is safe and healthy.** I need to check on everyone regularly and get worried if I can't reach someone. I'm often late for events or to meet friends because someone needs me or I need to check up on someone. But it's important that I make sure everyone is okay before I tend to myself.

- **I must be sure that things always turn out the way I need them to.** I decide my outcomes before they happen because I know how everything "should" be. It's important to me that I plan what should happen and strive to make it so. Even if I have to work harder or give up doing something I looked forward to, I will see everything through to the end I chose. I am a slave of getting the right results.

- **I must be sure about everything all the time and leave no stone unturned.** I don't like uncertainty. Actually, it makes me crazy. If I'm traveling, I double and triple check and reconfirm my flights and hotel reservations since I don't like surprises. I do everything I can to stay on top of what's going on in my life.

- **I can't disappoint anyone ever—otherwise I will be abandoned.** I need people to like me. The thought of being alone is scary. That's why I make sure to please everyone as much as I can. I loan money to people who I know are irresponsible and won't repay it but don't know how to stop. I hate that I don't get it back, but they need to know I'm there for them. I stop what I'm doing if someone needs me. Then hopefully they'll stick around. When I feel needed by others, I feel safe.

- **If I let go, bad things will happen and I will lose everything.** I feel like a tightly wound yo-yo, bouncing between my obligations and needing to stay in control of everything. It's stressful but necessary because, if I relax, things may not go as I need them to.

EXAMPLES OF THOUGHTS THAT IGNORE THE CONCEPT OF ESTABLISHING A SENSE OF ACCOUNTABILITY AND CAUSE ANXIETY

- **I am hopeless; I can never change.** I've been this way for my whole life and can't imagine that I could think differently. I've tried to tell myself that many of my thoughts and choices need to change, but it never works for me. I accept that I'm just an anxious person and must live with that.

- **It's unacceptable for me to ever have negative feelings.** When I feel a negative thought coming on, I push it right back down. I guess those feelings are still inside me, but I try to ignore them and certainly don't let others see me feeling bad about something.

- **Everyone is always against me all the time.** I feel like people don't like me for me. My family often lets me know what's wrong with me. At work it feels like everyone is competing with me because I work harder and put in longer hours. I get little support but lots of criticism. I feel unappreciated all the time.

- **No matter what I do, bad things always seem to happen to me.** I try so hard but still sometimes make mistakes or say the wrong thing. Often, circumstances that I counted on fall apart. I believe that I was born under an unlucky star because things rarely seem to go my way. So I kind of expect things to go wrong, and they do.

- **If I reveal my true self and act authentically in any relationship, people will discover who I really am and reject me.** I don't like me or think I'm good enough for most things I want. I keep my true personality, opinions and desires to myself and try to be the kind of person that people have come to expect me to be. Sometimes I feel like an actor, but that persona keeps people around. Maybe some people would be okay with the real me, but I'm too scared to risk it.

- **What's the point of doing anything at all if I know it's all going to end someday anyway?** Why bother to change myself? I'll be in the ground one day anyway. This is what I'm used to. Things end for me. The company I liked working for downsized and I lost my job. My relationships always end, even when I'm happy with them. Why bother finding something else I like that I'll eventually lose?

It took me a long time to identify some of my own negative thoughts and how they created my reality. When I got upset, I used to subscribe to the fatalistic outlook that "you work hard all your life and then you die." This misguided outlook temporarily absolved me from taking responsibility for changing the quality of my life. It prevented me from holding *myself* accountable for it. It seemed easier to put the responsibility for the problems in my life on to other people, or God, or the universe or whomever I was angry at. It took me a long time to dig deep and recognize that I had to take responsibility for believing that "you work hard all your life and then you die." It allowed me to slowly accept that it didn't have to be that way.

HOW TO REFRAME YOUR THOUGHTS

Once you've identified your negative thoughts, the next step is to replace them with more balanced and realistic ones. This doesn't mean you should adopt cheap, rose-colored affirmations that have no substance. Instead, create responses that challenge your negative thoughts and offer an alternative way of thinking—ones that are rooted in your own personal value system that you yourself create from the reality you'd like to live with. Take one identified negative thought that feels major to you from each of the four concepts—clinging to consensus reality, thinking with a dualistic mind, having a need for control and not establishing accountability—and break them down to more balanced ways to view them.

I had to learn how to reframe the beliefs ingrained in me when I was a child. My father's message as my role model was that a responsible adult must worry about all things and must stay vigilant. If you let your guard down even for a day, all hell will break loose. Because I trusted him, I developed a belief that to be safe in the world, I also had to worry all the time and be in control of everything, even if I had nothing to worry about. Sometimes, when I was not worrying about anything, I even worried that I was not worrying as my father told me to do, and it made me feel guilty.

My reframe came after I realized how much this tightly wound grip on my life was scaring me even more. I learned that letting go, little by little, of the tenuous grasp I had on my life wouldn't necessarily prove disastrous. It also helped me to understand that this was a fear-based value programmed in me by my father many years ago. This was *not* my value system. It was *not* my fear. It was his. The truth was I had only to focus on the things that I *did* have control over and most important, to establish what being a responsible adult meant to me and my own personal value system. Once I did that, worrying did not even make the list.

In the following section, I've broken down some of the common beliefs from the previous section to show you how you can reframe those thoughts into more realistic ones that are less likely to produce anxiety. Look for the keywords that make the beliefs unrealistic and replace them with words that feel more comfortable.

"I should be more productive/creative/ambitious."

This is an exceptionally irrational statement that reflects a consensus reality belief, but many of us still say it to ourselves on a regular basis. As discussed in Chapter 3, the keyword that causes the strongest negative reaction in this unbalanced statement is *should*. If you make a statement like this, you indicate a belief that there's some written law floating around in the ether that dictates how productive, creative and ambitious you should be, which of course there isn't.

Think about all the "shoulds" you adhere to. You "should" according to whom? Your parents? Society? Your religion? Who's driving the bus here? You are. This "should" thinking is a consensus reality type of belief. Once you've acknowledged having a thought based on a consensus reality belief, you can restructure the statement to fit into your own definition of what productivity, creativity and being ambitious actually means to you. With practice, you can learn to reframe any consensus reality statements you identify by using words that stabilize the extreme irrationality of the "should" basing them on thoughts that are more aligned with your personal ideas and values. An example of a reflective but not reactive reframe for "I should be more productive/creative/ambitious" might sound something like:

> I would prefer to be more productive/creative/ambitious in my life. But first I have to figure what "more productive/creative/ambitious" actually means for me. I will contemplate how I, myself, can measure that goal and see what steps I can take to gradually reach it.

This is clearly more balanced because it allows you to be the author of what "being productive" means to you, with clear boundaries and limits. This also dismantles the abstract authority of the "should" by turning the responsibility on to yourself. *You* define what you're going after to suit *your* needs and abilities. And "I would prefer" is much more empowering than "I should." And it's also much less stressful!

"If I make a wrong decision about anything I do, it means I am stupid."

This is a classic dualistic mind statement that will be very debilitating if you don't challenge it immediately with a reflective attitude. It leaves you wondering about what the "right" decision is, and infers that not choosing the "right" decision is somehow immoral, which is far from true. We make decisions based on the information we have at the time. Hence there is no right or wrong.

The keywords that can stir up the strongest reaction are *anything I do*. The word *anything* is one of those absolute words I discussed in Chapter 3, which doesn't have a place in the natural scheme of life. As you know, life is full of varying degrees of circumstances and has a subtle balance, so no one can *ever* be correct about anything all the time. In addition, the word *stupid* also goes along with a dualistic mindset because it unfairly labels you according to the false standard imposed by the concept of black-or-white extremes—believing you can only be either smart or stupid.

Reframe the statement to balance the all-or-nothing thoughts with words that are more aligned with *your* personal ideas and values:

Life is actually filled with subtle balance and varying degrees. There will be times when decisions I make won't work out for me but it doesn't change who I am. In the future, I hope to make decisions in my life that are compatible with what I believe to be appropriate for me and my best interests and know that I'm doing my best.

This reframe is less rigid and more kind and compassionate than the dictatorship of needing to be right about every decision you make. It forces you to reevaluate the black-and-white nature of your thinking, empowers you to take action against your tendency to go to extremes and reminds you about the importance of paying attention to the abundant gray areas that make up most of our lives.

"If I let go of control, bad things will happen and I will lose everything."

This statement is a clear example of a distorted way of thinking that produces anxiety. Yet it's a negative belief that many people adhere to in hopes of feeling more secure. As mentioned in Chapter 5, this type of negative thought means you assume that grabbing on to life with a relentless grip is the only way to live safely. In reality, you can only have just a little control over most things in your life, and in order to gain some real control, you must actually let go of some of it first.

I will lose everything is another example of an unmeasured, unfounded, catastrophic absolute that's very unrealistic. If you think about it, losing "everything" is simply too wide a generalization to have a specific meaning. It has no true validity whatsoever and is an illusion of control. Restructure the statement to reflect a more rational attitude that will be more in line with what you *do*, in fact, have control over and accept what you have no control over:

> *Trying to achieve control of everything is an illusion. I will instead assess the things I do have control over in my life and focus on them. I accept that letting go of control in certain important areas of my life will be scary. But in the long run, it will reduce my anxiety, which is good.*

"No matter what I do, bad things always seem to happen to me."

This statement is another example of unfounded thinking that's gone overboard. It's a disabling negative thought that renders you helpless by placing the responsibility for your happiness outside of your reach. It victimizes you into believing that your ability to change your situation is hopeless and you'll always have to live with negatives. As mentioned earlier, if you blame others and the circumstances for your lack of happiness or success, you will consistently feel vulnerable and sometimes even incapacitated.

The words in this particular negative thought that cause the most damage to you are *no matter what I do* and *always*. *No matter what I do* is a blanket phrase that overgeneralizes specific things in life that truly didn't go well for you. To project that bad things will definitely continue to happen for the rest of your life based on some past incidents is irrational. This self-imposed impotence—the belief in your inability to keep things from going wrong in the future—will cause you to experience great mental angst. In this state of mind, you're convinced that you're unable to exercise your free will to choose your destiny.

By reinforcing this belief, you surrender to feeling doomed about your life and assume that any efforts you make won't pan out. You'll blame everyone and everything outside of you for never being able to turn your life around and have good things happen.

Reflect a more rational approach that begins to take back some of your power:

> *Bad things happen every day. When I blame the world for my problems, I become disempowered. Since I am the author of my destiny, I will instead empower myself by creating a future of personal accountability where I hold myself responsible for my happiness. In doing so, I can create more happiness and enjoy having less anxiety.*

This reframe encourages you to take action and choose personal accountability over being a victim. It also acknowledges that life is indeed difficult and that unfortunate things can happen any time. But in that acceptance is the option to do something about it, which can lower anxiety. That gives you more control than any of these beliefs can provide.

Learning to reframe your thoughts as I did in this chapter can take time, but it's worth the effort. It will probably feel uncomfortable or even scary at first, so go slowly. Don't get angry with yourself if you have trouble with some thoughts that you've had for a long time. You don't want to add to your anxiety as you try to temper it. Just keep trying.

Chapter 8

SOLIDIFYING YOUR SENSE OF SELF TO BUILD A HEALTHY SELF-ESTEEM

*"Your vision will become clear
only when you look into your heart.
Those who look outside dream,
those who look inside awaken."*

—Carl Jung

Now is a good time to awaken your stripped-down sense of self that's beginning to free itself from falling victim to the automatic thought patterns that can exacerbate anxiety. Ultimately, you can never fully rid yourself of the fixed beliefs that you may have been raised with. But you can learn to create enough space between your unconscious thoughts and the ones they produce so you can keep the fixed beliefs outside of your own perception of yourself and the world around you.

FINDING YOUR REAL SELF

The first phase of getting in touch with your real self is to step outside of your public persona and think about what aspects about yourself that you like.

This can be hard to do if you're used to identifying with old beliefs. Doing this exercise will not be easy, but it's necessary. Neale Donald Walsch says, "Your life begins at the end of your comfort zone." You'll probably feel quite uncomfortable in the beginning. But it comes with the reward of helping to ease your anxiety.

When I give this assignment to my patients, many report that it brings up uneasy feelings because focusing directly and deliberately on themselves feels overly indulgent. Feelings of shame often come up from the many years of deflecting personal attention. In many cultures it's the norm to put oneself second to the needs of others and to think that you're part of a whole instead of a separate person.

Try to be as specific as possible about what you recognize about yourself that feels good. Describe why you find each quality likeable.

For example, do you like that you're structured and almost methodical in your thinking, even though others make fun of it? Or maybe you're the one who is too critical of it despite the fact that it still brings you results. Do you like a particular hobby or personal interest that gives you joy and lifts your spirits, even though others disapprove of it at times? Perhaps you join in that disapproval because others do? Do you like and possibly dare to admire a personality characteristic about yourself, such as having a good sense of humor or that you can be dark and mysterious?

Do you like a specific physical aspect of yourself—for instance, your thick head of hair, or your curves, or your trim physique or the color of your eyes—but you're too ashamed to appreciate anything like this because you don't want to seem vain or pretentious? That belief can prevent you from owning the wonderful qualities you have, and you do have them. We're often so worried about what people will think if we're not modest or even self-deprecating that it can be easier to forget about our assets. Feeling pride in yourself is not the same as being vain or conceited. Yet you might have been brought up to believe that. This consensus belief can keep your self-esteem low if you're scared to recognize the beauty and abilities that you possess.

In the spirit of venturing out of that comfort zone of needing to be overly modest and show great humility, try the exercise that follows once a day. You can do it regardless of whether your hesitation to recognize your good qualities was imposed on you or whether you imposed it on yourself to avoid the risk of alienating anyone by being judged as vain or as a braggart. Make sure your answers reflect what *you* think, not what others have said about you. Keep a notepad handy or create a document on your computer that no one else can see, and leave yourself a note or other reminder so that you remember to do the exercise daily.

EXERCISE
WRITE DOWN THREE THINGS YOU LIKE ABOUT YOU

Every morning write down three things you like about yourself and why.

For example:

I like my wicked sense of humor.

Because it helps me to laugh at myself and not take life so seriously, it helps me find the absurdity in even the darkest of things and it cheers up other people.

I like

Because

I like

Because

I like

Because

The next exercise will begin the process of solidifying your sense of self by identifying three of your strengths. As with the last exercise, they can't be strengths that someone has complimented you on or pointed out to you. They must be three strengths—things you do well—that you recognize and believe are actually true. This can also be hard to do because it may push buttons that are related to your old habits, stirring up feelings of shame and general discomfort as you think of things, which is natural. Remember, you're not used to singling out yourself in this way and you probably haven't evaluated yourself in terms of the positive features you have.

Doing an inventory of your personal assets or strengths in whatever capacity you have them is key to building a good sense of self-worth, which eventually leads to a more healthy sense of entitlement. If you can begin to acknowledge yourself in the light of ability and human value, you'll start to know and respect yourself better. You never may have thought about your strengths because you were too focused on doing things the way they're "supposed to be" done or being down on yourself for your weaknesses. When you go through life on auto-pilot that's been programmed by others, you may take yourself and your assets for granted. Now it's time to acknowledge them.

As in the first exercise, try to be specific about exactly what it is that you're good at. For example, you may give it some thought and begrudgingly acknowledge that in crisis situations you're efficient and think fast on your feet—most of the time a strength like this is not recognized as one because you may brush it off as somehow not counting and take it for granted. Or maybe you're very responsible and reliable in your relationships or in your career. Or you may be acutely sensitive and compassionate about other people's feelings, or perhaps you're a good listener. You may be a good writer but have been too self-conscious to use your skills or to show what you've written to anyone, so you don't acknowledge this gift.

Do you have a way with words when you speak, or possess creative abilities that you never let surface, or have another skill that you keep dormant because you're too busy comparing yourself to other people? Remind yourself of Michelle Charbonneau's words, "When we compare, we despair." Since you're a unique individual, there can't be a comparison between you and other people. There's room for all good abilities. There are many good writers and musicians and accountants and people doing the things you know you're good at, even if you squelch your urge to use these abilities or let yourself feel pride about what you can do. So dig deep to find your strengths.

Try to think of something you did in the last week or the past year that demonstrated a capacity or skill in something. Or think of a time when you helped someone with a task. It can be anything; nothing is too small. Actually, the smaller the accomplishment the better. It's unrealistic to expect to do great things all the time. You're likely to be disappointed if that was your goal. But how wonderful it would be if you could do some little things really well at least some of the time. Small things, like helping your child with homework or supporting a friend or coworker after a stressful experience, are easier to achieve, but they also add up. Each one can make you feel better about yourself. Try this exercise once a day. See what happens.

EXERCISE
WRITE DOWN THREE STRENGTHS

Write down three skills, abilities or talents you possess and their positive attributes.

For example:

I am very good at listening attentively to others and conveying compassion.

Because I can see that when I do listen to others and convey kindness, they trust that I will not judge them. It also makes me feel good that I can be there for someone else, and it reminds me of how I, myself, would like to be treated as well.

I am very good at

Because

I am very good at

Because

I am very good at

Because

THE POWER OF AUTONOMY

Being able to identify and gradually approve of your separateness from the rest of the world is an important key to solidifying your sense of self. You'll embrace that you're a distinct human being—an individual among other individuals instead of trying to be what others think you should be—without the dualistic mindset of seeing who you are as good or bad and right or wrong. As you practice recognizing your individuality by doing the preceding exercises to identify your personal likes and strengths on a daily basis, without judging what you find, you'll begin to see changes in your view of the true individual you're becoming, and you may start to see your symptoms of anxiety subside. Be patient and keep building your self-appreciation, and the benefits will eventually come.

It's important to understand that the term *separateness* does not mean *indifference* or *isolation from others*. The power of separateness lies in the autonomy you develop that also forms solid connections. But it's important to value the treasured spaces that exist between you and the people you're close to. That space or boundary that distinguishes you from everyone else is what the existentialists refer to as one of your basic human drives—"to strive for identity in relationship to others." If you don't strive for this autonomous identity, you're likely to allow that space to be compromised, and you'll begin to lose yourself again.

When I first began to establish my personal boundaries with others, including intimate relationships, I was initially hesitant and feared rejection. For most of my life, I was a passive communicator, especially with my family. Hence, I was anxious all the time because I had no sense of self to rely on. However, once I began practicing asserting my autonomy and setting personal boundaries, my anxiety decreased and things started to change. I noticed that people respected me more and didn't reject me. Most important, I began to appreciate my value as a separate person, and my self-esteem grew. The power you gain from maintaining an identity that's separate from others lies in your ability to stay intellectually differentiated from their ideas by adopting only those that fit into your own way of thinking. It's also reflected in your ability to stand alone amidst any chaos that may be around you. When things get a little heated and there's conflict that puts that space in jeopardy, you learn how to hold that centeredness and stick to your guns without having to fire them. Patience is required here, as this power or the choices you make to carve out your space in relation to the people you care about may not be simple and can require a great deal of courage.

It can feel like a daunting task to alter how you relate to people as you begin defining personal boundaries, especially with family members. It can also bring up old fears of getting disapproval from them or being abandoned. For example, if in the past, family members have become upset with your attempts to assert yourself as an individual by having opinions they disagree with and developing a voice that's different from the family's, your reflex to remain passive might pop up to avoid conflict. To assist in harnessing this power of being your own person, you must figure out how to communicate this to others in respectful and kind ways. The boundary— the line of distinction that you must draw between what you will and won't accept—that you use with people may appear to many as being uncaring or cold. That's why it's good to do your best to introduce this new attitude gently, in a nonthreatening way.

One of the best ways to communicate your choice to be an individual who thinks for himself or herself is to use an assertive style of communication that's boldly firm but civilized, friendly and courteous.

Assertive communication is a critical skill in the foundation of drawing a clear line in the sand that says that you are richly diverse and wonderfully complex. Next, I'll share one story about a person who learned to do this, and discuss in more detail assertive communication and how it differs from the other two main types of communication: aggressive communication and passive communication.

TINA'S STORY

Tina was an executive assistant for a large corporation. She came to me for treatment of her workplace anxiety. Over time we discovered that much of her stress came from her inability to assert herself. She believed that if she stayed passive in all her interactions with coworkers, she'd be safer and not lose her job. Because of this, she reported that there were a few coworkers who often took advantage of her and sometimes coerced her into doing more work than she was paid to do. She knew this was unfair but was reluctant to do anything about it.

Tina said that everyone seemed to like her and people always told her she was the "nicest person ever." But she learned over the years that while there were benefits from being passive and never saying no, she felt unfulfilled and often didn't get what she wanted. Tina aspired to being a lot more than an executive assistant but didn't know how to assert her goals. Tina's self-esteem was at an all-time low when she came to see me. She felt stuck and didn't think she had much value. I placed her on a steady diet of taking daily personal inventories of what she liked about herself and her strengths. I also encouraged her to practice honestly expressing her thoughts and desires more with others instead of defaulting to being passive, including saying no instead of always being acquiescent.

Eventually, Tina started to respond well to her personal exercises, and her self-worth began to seem more real to her. The exercises helped her to believe in herself based on *her* own likes and the strengths that she identified. The fact that she alone came up with her own personal asset list was a game changer for her. After several months of working the exercises and practicing letting go of passivity, she reported a decrease in her anxiety and began looking for new job opportunities that were more in line with her career goals. Tina transformed into a woman with a mission and this time, she had herself to rely on. Tina was finally aligned with the real Tina instead of the Tina people expected her to be.

COMMUNICATION STYLES IN ACTION

Your style of communication influences the response you'll get. Be aware of what you use and consider adopting a more effective one. Here are some common styles.

Aggressive Communication

When you speak with an aggressive style of communication, you may feel like you're standing up for yourself, your feelings and your beliefs, and drawing that important boundary line of individuality. But this may not be a style that delivers your message in a courteous and respectful manner. If your message is perceived with negative overtones or the delivery puts people off, then the point or the boundary you're trying to get across may not be understood or even heard.

Aggressive communication isn't usually well received and tends to create a defensive attitude in the other person, which affects how your message is heard. It may make the person want to get back at you or tune you out and not hear a word you say with objectivity. Aggressive communication tends to make people feel directly violated, in a sense, because it's an intimidating style that often steamrolls them by using a loud voice, screaming, demands, manipulation, humiliation, blaming or control. Scare tactics such as hostile body postures and even facial expressions can also contribute to this aggressive type of communication that's not conducive to establishing boundaries in a healthy way.

Because of how it may be received, aggressive communication is not an effective way to get what you want. You might alienate people instead. Remember, without healthy interpersonal boundaries that identify your individuality and help you deepen your relationships, you won't have a solid sense of self. Aggressive communication doesn't foster profound and trusting relationships. Instead, it creates hostile and conflicting dynamics that leave others feeling scared and manipulated. It also places the burden of the problem on others and doesn't let you take responsibility for your life. Using aggressive techniques is not a healthy way to show respect.

Passive Communication

In contrast to an aggressive style, passive communication withholds thoughts, feelings and beliefs that may seem like you're playing it cool and trying not to rock the boat. But it doesn't foster healthy interpersonal relationships any better than aggressive styles do. When you communicate passively, you tend to relate to others dishonestly or apologetically because you don't express what you really feel in an effort to avoid getting a negative response. In a sense, you violate your own rights by not being true to what you think and feel. In addition, you not only deprive yourself of the right to your individuality within a relationship but you also deprive others of knowing what you really think. This passive stance leaves relationships lacking in trust, honesty and authenticity because others don't know what you truly feel.

Using this style, your chance of getting what you want is even smaller than when you relate aggressively. Passive communication does not contribute to creating a solid sense of self because you're too afraid to draw any visible line that separates you from others. This low-key style of communication is not necessarily caused by shyness. It comes more from being scared of looking different in people's eyes and an aversion to feeling unique, which is the cornerstone of identifying yourself and reducing anxiety in the long run. It keeps you stuck living in the reflection of how others want you to be.

When you're passive in a relationship and don't define who you are within it, you're not living authentically and won't be able to create your own reality. Passive communication forces you to live in the reality that other people dictate. Deep down inside, you may want no part of it, but you go along with that reality rather than express how you really feel. It also tends to place the burden of a problem or conflict entirely on you if your lack of clear communication naively lets others off the hook.

Assertive Communication

Assertive communication is a more suitable way to express yourself in an honest, authentic and nonviolating manner. Like aggressive and passive communication, it doesn't guarantee that you'll get what you want, but it's the best shot you have. Assertive communication is more about verbalizing how you truly think and feel, openly and honestly. It's the foundation of acknowledging that "I am me" and "you are you." It expresses that "We are not one; we are not alike." Each person has his or her uniqueness, and you're choosing to honor yours. And it's okay to be different.

Assertive communication helps raise your self-esteem by establishing your sense of self in any situation. It's the appropriate way to draw a boundary line that doesn't involve withholding thoughts and feelings or expressing them inappropriately. This direct communication is not about power, manipulation or control, nor is it about raising the volume of your voice or using intimidating body language. It is thoughtful and considerate and adheres to a process orientation instead of a results orientation. It's more about how you are with others rather than trying to get what you want from them.

Assertive communication is based on "I" statements that are used as a way to take full responsibility for your feelings, thoughts and actions. The "I" statement is a neutral stance that does not blame or point fingers at others. "I" statements are specific. They don't use absolute language, such as *always, never, should* or *forever* since these assume impractical expectations of others and indicate an unrealistic view of the world, which usually isn't true. Absolutes have no place in healthy human interactions, especially during conflicts, since they typically make the other person angry or resentful.

For example, if you're arguing with your partner and feel like you're not being heard, you may say in desperation, "You never listen to me. You always disrespect me." There's a very good chance that your partner will become irritated by those words—*never* and *always*—because in all likelihood they do try to listen when they can. And, more important, it's hurtful to hear that you feel that way. The argument could then easily escalate. This won't establish healthy autonomy between you. The statement also uses *you,* placing the responsibility for how you feel onto your partner. A more appropriate way to express not feeling heard that better establishes the autonomy you seek is something like, "I feel hurt because it feels like you haven't really been listening to what I'm saying. I'd appreciate your paying more attention to my words."

Assertive communication has three parts:

What I feel: The first part is identifying specifically how you feel in this situation. It helps to make your exchange less harsh and allows the other person to see how and what he or she says or does negatively affects you. It also softens the conflict by introducing your subjective experience and communicating it with courtesy. Stating how you feel to others is significant because it helps you begin to establish the importance of who you are by acknowledging yourself and understanding your feelings. Other people can't do that for you.

What I see: This part simply identifies the behavior nonjudgmentally. Doing this helps you take responsibility for what's going on by not making the person you're addressing the cause of your pain. Instead, you focus on how their behavior causes you to feel a certain way. By approaching the issue this way, you're not directing the negative feelings you have at the person, so it doesn't come across as a personal attack. You're just sharing how their words or actions made you feel.

What I would like: This part identifies what you're asking for in the spirit of being an autonomous person and reinforces your personal individuality. This doesn't mean that you'll get what you want, but what you say will be received better. For example, using wording like *prefer* or *I'd appreciate* prevents you from falling into the bad habit of using absolute words to tell people things like, "You *should* listen to me better."

For example:

Step 1. What I feel: *I feel hurt.*
Step 2. What I see: *That you haven't really been listening to what I'm saying.*
Step 3. What I would like: *I'd prefer you to pay more attention to my words.*

To help you understand the different communication styles and see why assertive communication works best, I'll take a common interpersonal relationship issue that can trigger a lot of anxiety and frame it under all three styles of communication. Let's say your father is continually critical of you for choices you've made in your life. At a Thanksgiving dinner, with many of your relatives at the table, he sarcastically infers that he's disappointed in you because you didn't follow him into the family business and instead chose your own career path. Here are the three types of communication you can use to respond:

Aggressive communication: You would say something like, "You are so selfish, Dad! All you ever do is think of yourself. Why do you always have to ruin dinner with your petty grievances about what I choose to do with my career? You never cared about me anyway, so why pretend like you do now? Go to hell!" This kind of communication will ruin the dinner, and your father will probably respond in a defensive way that will escalate the tension between you. And you'll feel angrier when it doesn't get resolved since it's unlikely that he'll apologize after what you said.

Passive communication: You would offer no verbal response. You're angry inside but too afraid to express it. You shut down and quietly churn in your contempt. You then cower to him, bow your head, feel shame in front of your relatives, remain silent for the rest of the dinner and feel responsible for your father's pain. You also stifle anger at him. Or you get up from the table, walk away and withdraw.

Assertive communication: You would say something like, "I am very upset [what I feel] that you indirectly brought this up again with me [what I see]. I don't appreciate this right now. I am sorry that you're angry with me for following my dreams and not following yours. I would prefer [what I would like] if in the future, you would remember that there is a time and place for these types of discussions, and this is not one of them. So, let's change the subject [what I would like]."

Remember, assertive communication within your interpersonal relationships is not intended to change others nor to necessarily get the outcome you'd like. If any of these things do occur as a result of good communication, you're lucky. The truth is, assertive communication is intended for you—yourself. It can help you evolve as a human being, raise your self-esteem and promote your dignity as a unique individual. And in time, as you integrate speaking assertively into your everyday life, it can decrease your symptoms of anxiety as well.

COLIN'S STORY

I treated Colin for generalized anxiety for several months. Although he was a kind and harmless man, he tended to relate to others with an aggressive style. Through our work together, we uncovered that Colin, who suffered from very low self-esteem, was afraid to act warmly to others in general because he didn't want them to see him as a weak person. To him, a passive or gentle person was someone who would not be taken seriously. He also believed that to survive in the world, one had to be a little harsh. So to overcompensate for feeling scared and mostly inadequate inside, he adopted an aggressive style to appear strong and confident.

He didn't like that his aggressive style made him unable to form lasting relationships, which left him feeling very isolated and even more afraid and anxious about the world. Colin wouldn't allow himself to be authentic. He didn't give himself a chance to be liked for being Colin. He felt compelled to live up to a fake persona to avoid rejection. When he came to see me, he was in limbo—he was afraid that if he softened his approach, people would walk all over him, yet he was afraid of continuing to live in fear as the aggressor who was always on guard. In the end, Colin couldn't hide from the truth—his aggressive communication style left him feeling unfulfilled (he rarely got his needs met), misunderstood and feared by others.

He also realized that people lost respect for him because he tried to gain it from them the wrong way. Eventually, Colin learned that using assertive communication permitted him to be warm and gentle with people without appearing inferior. He could be direct and honest without having to blow up and be emotional. This allowed him to feel strong and important. He was able to deepen his relationships in both the workplace and with his family. It also decreased his anxiety in the long run because he understood that there was very little to fear by being authentic. People were going to like him or not. But he had a better chance of people liking him as an assertive man than by being aggressive.

BUILDING SELF-ESTEEM BY REDISCOVERING CORE VALUES

Another vital aspect in the process of solidifying your sense of self is getting in touch with your personal core values and, most important, learning how to stay connected to them. Whether you know it or not, you possess core values—things that are important to you—that drive your existence. These values are the pillars that support the infrastructure of your life and the reason why you get up in the morning. They're also the fabric of who you are as an individual, because they give you meaning and purpose. The core values I refer to aren't necessarily your parents' values. They're your own personal ones, like the ones I discussed earlier, that you've developed independently. There may be a vast difference between yours and your family's. It's up to you to review what you think are your core values and ask yourself if each one is aligned with what you truly want to live by. Through the years—in varying degrees—your core values tend to fall by the wayside, especially if you've become depressed or preoccupied with excessive worry over life's unavoidable difficulties.

As a result of letting go of core values, your self-esteem can take a big hit and plummet because you've lost your sense of direction. Without that direction and purpose, you can begin to forget who you are. Realigning with these values will give you insight as to where you can put your energy and what to begin focusing on as part of the process of rebuilding your self-esteem. When you rediscover your core values and make a conscious decision to live by them as best you can, you'll gradually begin to see changes in your life. And, over time, you'll start to feel better because you're in harmony with yourself.

What follows is an exercise to help you reconnect with your values. When I ask my patients to do this—similar to their experience with the personal likes and strengths exercises presented earlier in this chapter—many report that it brings up uncomfortable feelings because the direct focus on themselves again feels too egocentric. But remember, to build self-esteem you must identify and acknowledge that you are a separate individual in relation to others. If you can appreciate your uniqueness and value as a person, you may be able to appreciate those qualities in others, too.

EXERCISE
REDISCOVERING CORE VALUES

The following is a list of possible life values to help you get started with identifying your own personal core values that are important to you.

The idea is to get as specific as possible. Material things are also not the kinds of values you should look for in yourself. Therefore, things like money, retirement plans, real estate, cars and other monetary-based values are not considered to be core values.

Use these as a guideline, but come up with your own, too:

__ Commitment to family	__ Commitment to spouse/partner
__ Commitment to community	__ Commitment to God
__ Spirituality	__ Health
__ Nutrition	__ Exercise
__ Integrity	__ Responsibility

___ Self-respect   ___ Honesty

___ Self-reliance   ___ Sense of humor

After you've compiled your list think about what you are willing to commit to doing each day that are in accord with these values. For example, if one of your identified core values is a sense of integrity, to align your behavior with that value, you may decide to make amends with someone you had a falling out with in the past. You may call a family member and perhaps open up a dialogue about an issue that's unresolved between you using assertive communication. Or you may choose to follow through on a task or a goal you've put off for a while that's been eating away at you and making you feel inadequate for not getting it done.

If you identify with having spirituality as one of your core values, choose to engage in some mindfulness meditation before or after work. Or you may choose to attend services at a place of worship, or you may pick up reading materials that inspire and reconnect you to whatever your higher power is. Or you may decide to spend time experiencing nature—walking in a park, strolling on the beach or hiking in the forest. Or you may just choose to sit somewhere quietly during your lunch break and take in the sights around you. You'll know what to do when you reconnect with your value.

In the exercise that follows, record your core values and then list some things you can do now.

For example:

Core value #1—Spirituality or connecting to your higher power

1. I will do mindfulness activities and/or meditate every morning for 15 to 20 minutes before I go to work.
2. I will attend a place of worship for services once a week and while I'm there, I'll have a conversation with at least one or two new people.
3. I will do 30 minutes of mindfulness walking outdoors.

EXERCISE
CORE VALUES AND A LIST OF ACTIONS/ACTIVITIES

In a notebook or on your computer, make a list of planned actions/activities you can schedule or commit to once a day.

Core value #1

Actions/activities I will take today:

1.

2.

3.

Core value #2

Actions/activities I will take today:

1.

2.

3.

If you do this exercise once a day every day for one month, you're likely to experience a change or a shift in your thinking about yourself and your place in the world.

Chapter 9

EXTREME ACCEPTANCE

The X Games for the Mind

*"The greatest wisdom is to make
the enjoyment of the present
the supreme object of life
because that is the only reality,
all else being the play of thought."*

—Arthur Schopenhauer

The concept of extreme acceptance is purely a state of mind. It's a sweeping mental attitude that in many ways can act as a universal coping skill. Extreme acceptance is not a destination or a goal that you reach and then you have it. It's an ongoing process of exercising your mind to be flexible and tolerant of anything that life throws at you. The mindfulness of extreme acceptance encompasses all four of the concepts discussed in previous chapters and finding acceptance with them all: challenging consensus reality, balancing the dualistic mind, letting go of the illusion of control and establishing an accountable self.

FLEXING YOUR MIND

When you learn to exercise extreme acceptance for all aspects of your life, you will gradually shed any previously assumed negative belief systems you might have had that you felt required to live by. Then you can open up a new world of thinking and responding. Applying extreme acceptance to your life means that you must dig deep inside of yourself and accept what may seem outrageous or what goes against everything you've been taught.

One of the hardest things I had to accept in my life, especially when I first started doing this work, was to accept that I could not please everyone all the time. Later, I also learned that I could not even please everyone *most* of the time. Plus, "everyone" and "all the time" were irrational absolutes that were driving me crazy, though I didn't know it at the time. When I first began to accept my powerlessness to please everyone, my anxiety spiked as a result of this mind shift. It felt outrageous. I had been a people pleaser for years and it was the only way I knew how to relate to others. Through extreme acceptance, I gradually let go and accepted the bitter truth: that every now and then I was going to either disappoint someone or someone was not going to like me. Once I grasped that concept and accepted it, I felt better. The pressure was off.

Practicing extreme acceptance can be scary at times, because if you have any drastic change or abrupt shift in what's familiar to you, there's an emotional risk involved. But despite the potential risks and high stakes, Eckhart Tolle reminds us: "Acceptance of the unacceptable is the greatest source of grace in this world." It's not easy to accept what might seem unacceptable to you right now, but the more you can accept what's going on in your life, the more you can relax about it. That in itself is a great way to reduce anxiety.

WHAT IS ACCEPTABLE?

It's important to be clear that extreme acceptance is *not* indifference, apathy or blind resignation. It doesn't involve skirting personal responsibility and it's certainly not related to giving up or quitting. In addition, it doesn't mean that you must view whatever misfortunes you may have suffered in the past as a good thing. And it also does not imply that you should arbitrarily forgive wrongdoers and absolve them of culpability. Although this is a personal choice, these actions and attitudes may not always be healthy for you. However, every case is different and relative to the situation.

Extreme acceptance focuses on more positive aspects of life. It's a coping mechanism to help you stop fighting reality and learn to roll with it more. When you can accept the things you cannot change, you can put more energy into controlling your response to the world instead of letting what's going on create anxiety. When you have extreme acceptance you'll experience the following:

You accept things as they are, without judging them.

Whatever comes your way, you mindfully refrain from overanalyzing and possibly judging what it means to you or why it's wrong or bad. You accept that it is what it is. Whatever happens—happens—and life goes on. When you have extreme acceptance, your patience increases, which allows the natural flow of what's going on around you to take place. Keeping this perspective means letting go of the innate reflex to interpret all of your experiences with a critical eye and letting go of the ways you viewed life that you recognized in Chapters 3 through 7.

In addition, holding on to resentments and grudges that result from making judgments deprives you of gaining the insight and growth that can develop from the experience. Sustained resentments about things that happened in the past are like lazy forms of grieving because, as you probably know, it takes a great deal of inner strength to let go of anger and feel the pain. Anger, resentment and other negative emotions are unhealthy outlets that cover up the pain. It's better to let it go completely and move on.

For example, Edie was a patient who incessantly judged herself for the way she looked. She resented herself for having a curvy figure, for not having nicer hair, for having brown eyes instead of blue—the list went on. She was even highly critical of herself for succumbing to her symptoms of anxiety and coming to see me in therapy. She could not accept that she even needed some help. Most of the work I did with Edie centered on finding acceptance with who she was. She resisted that for many months, and at first thought it was fruitless to make such a radical suggestion.

But eventually Edie came to an understanding that it was better and easier to accept herself as she was than to wish she was somehow different. She learned that pain was the difference between reality and what she wanted reality to be. Once her acceptance of situations in her life kicked in, she began to feel better.

You accept others for who they are— and for who they are not.

Acceptance of other people who are different than you seems to be a very difficult task for many people, especially when you've been stuck in rigid patterns of thinking. But when you opt for extreme acceptance, you accept the differences between you and other people. And, more important, you're grateful for the genius of diversity in thought, personal values, sexual orientation, religion, points of view, race and other variations of those you encounter. Extreme acceptance doesn't assume expectations of others but instead attempts to adopt a compassionate acknowledgment of others without imposing your own prejudice.

Therefore, when you have extreme acceptance, you strive to view everyone, as well as yourself, as unique humans in constant transition. You also try to accept that all of us are complex, multilayered individuals who are highly vulnerable and superbly imperfect. Extreme acceptance is about cultivating tolerance. You don't have to agree with everyone's thoughts and actions or like them, but you accept their right to have their own ways, just as you have yours. When you have extreme acceptance you also understand that you don't have the power to change others, no matter how much you might like to. You know that you can encourage others, influence them, maybe even get some to see or accept your point of view on some level, but beyond that, you're powerless. That's why it's important to accept that it's best to take control of yourself and leave others to do the same for themselves. Then you can stop worrying about them.

For example, when thirty-two-year-old Jeremy was a child, he observed his parents immediately judging people based on personality and appearance characteristics. All people, even family, were critiqued and "sized up" by his conservative parents based on stereotypes and preconceptions that had little to do with true character and integrity. People who had tattoos or body piercings were deemed dirty and untrustworthy, regardless of who they were or what they had achieved in life. Those who had alternative lifestyles with differing values were also judged harshly. People who spoke up and challenged convention were seen as irreverent or subversive.

Even as an adult, Jeremy subconsciously assumed the same mindset and began seeing people with a scrutinizing eye. He split people into categories as if they were articles of clothing. As a result, he only had acquaintances in his life and no true friends. When we began, Jeremy didn't understand his lack of close people in his world but he finally recognized that if he judged others so superficially, he was probably judging himself that way, too. As the sessions passed in treatment, Jeremy slowly let go of his parents' rigid beliefs about people and accepted others as they were. That helped him to accept himself, too.

You accept that all things must end. Nothing is forever.

Extreme acceptance teaches you to adhere to the mindset that life is merely a series of beginnings and endings. All good things end and, thankfully, all bad things end, too. Remaining overly attached to people, careers, success, specific outcomes and earthly possessions causes suffering and increases anxiety. If you can learn to look at loss or the end of a relationship as simply another ending in your life that's a normal part of being human, you're likely to experience less distress over it. When you're not too emotionally committed to what happened, it's easier to accept it.

As overly simplistic as it may sound, Dr. Seuss had a great way of looking at loss and endings: "Don't cry because it's over, smile because it happened." The finite nature of life and the knowledge that even your own life will end someday can, in fact, be a motivating insight that can give your existence a great deal of meaning if you allow it to. Things begin and things end. The concept of death and dying in this instance can be a friend instead of something to fear. People will come and will go. That's life. When you can accept this, you'll have more control over your anxiety.

For example, Larry had a successful thirty-year career as a designer of office buildings, and he was fifty-eight years old when he faced early retirement after his architecture company was forced to downsize. For years, Larry's identity had been tied to his job. Without his work to define him, he imagined an empty and grim future. I remember him saying one day, "Without my job, I don't know who I am." Larry and I worked on his inability to accept that he was now moving into the next phase of his life, and there was very little he could do about it. Like many of us, he wanted to be young forever and have the ability to decide when he would stop working on *his* terms. He fought the idea of retirement for months. Eventually he began to accept that this ending could actually be a new beginning that may not be so bad.

After many sessions in therapy, Larry embraced his new beginning. He did it by accepting the inevitable and by appreciating all he had accomplished over the years, like the many buildings he had designed in cities throughout the country and the countless people he had helped and mentored. He realized that acceptance would help him traverse through his remaining years with integrity about what he accomplished despite being retired. The alternative was to live with desperation about the potential emptiness in his life. It was his choice to make. He chose acceptance.

You don't ask unanswerable questions.

When you have at least some extreme acceptance, you catch yourself from asking unanswerable questions like, "Why was I born into my particular family?" or "Why can't I be smarter, richer, more attractive or more creative?" If you try to answer these unanswerable questions, all you'll come up with are interpretations and theories that don't provide a satisfactory answer. Often there are no definitive or logical answers, and anxiously ruminating about why something turned out the way it did is a waste of time since you'll never find good answers for those kinds of questions. Letting them haunt your thoughts just creates anxiety since you can make getting the answer important but it will never come. Other examples of unanswerable questions are:

- Why do all people have to die one day?
- Why do I have a chronic illness?
- Why are there evil people in the world?
- Why is life so difficult sometimes?

In most cases, there are no viable answers to these questions. A more important question to ask is, "Why now?" It's better to look at what you're doing today that might be adding to the problem. Why does what you're asking matter so much to you now? What's going on that makes you ask these kinds of questions and dwell on wanting answers? Your here-and-now thinking is the only reality you can count on.

Petra learned this. She was a happily married mother of three who came to see me because she was anxious about her youngest child, Eliza, not being as good a student as her other children. Eliza was a creative type and interested in music. She played the piano, composed music and dreamed of joining a jazz band in the future. Eliza's academic skills were poor at best but she managed to pass all her classes with a C average. She was also very social and was one of the more popular girls in school. Her demeanor was positive, and she was always polite and kind to people.

Because Petra's older children were straight-A students, she could not appreciate any of Eliza's nonacademic assets. Petra couldn't accept her daughter for being a musician and not a scholar, and she suffered a great deal over trying to will Eliza into being different. She lost a lot of sleep and sometimes incapacitated herself with anxiety over this. It was destroying her life. She often asked in session, "Why is this happening to my family? Why us?" In our work together, Petra recognized that, in her mind, accepting that Eliza was not special academically meant that she didn't care about her future. She was also afraid it would make her seem inadequate as a mother. Consequently, she felt like a failure.

At first Petra fought me about the idea of accepting Eliza for who she was. She was incensed with me for even bringing it up. But her mind shifted into acceptance after she stopped asking, "Why?" and just let it be for a while. She soon reframed her thoughts—understanding that her daughter's happiness was more important than her academic skills. Most important, she decided it was time to begin to appreciate what Eliza *did* possess: a positive and healthy state of mind, excellent social skills and a love for music. Her daughter's passion for music and determination to be a great composer soon inspired her, and her anxiety left as she developed pride for Eliza. However, this epiphany would not have manifested to her unless she first accepted what she considered to be unacceptable. She had to practice being flexible despite her fear.

THE IMPORTANCE OF GRATITUDE

Extreme acceptance can't exist without bringing gratitude into the picture. In order to reconcile that you're making an effort to accept difficult people and situations without judgment, and to accept that all things will end someday, it's helpful to consciously feel grateful for all the good in your life now. This means taking stock of the basic truths you have right now so you can acknowledge and appreciate them.

Your mind runs on perpetual autopilot unless you consciously decide to change it. It's up to you to learn how to switch over to a manual mode and self-manage your thoughts in ways that serve you better. This process is not necessarily a destination or a goal. It's simply a daily practice, like brushing your teeth or tidying up your house. Managing your thoughts can become a habit when you make the effort. At first you'll have to monitor them so you can make adjustments. But eventually your new, healthier thoughts will be the ones that come to mind first. Feeling good from thinking about all the things you can be grateful for helps you change those habits.

Zen master Lin Chi said, "The miracle is not to walk on water but to walk on the earth." In accordance with that thought, the wonder of gratitude is to be gutsy enough to peacefully embrace the simple things you have every day and take notice of your many gifts. As you begin to strive for extreme acceptance, you can also begin to trust that gratitude can be a radical alternative to former methods that you've used for coping. The following exercises can help you to take your focus away from negative and irrational thoughts to having more sensible and grounded ones.

EXERCISE 1
CULTIVATING GRATITUDE: THINGS TO BE GRATEFUL FOR

Write down three things you're grateful for in your life today.
Complete this exercise every morning for one month. Be very specific about why you're grateful for each one you choose. Although some repetition is okay, try to look for new ones as much as possible when you do the exercise so your gratitude can expand.

For example:
You might be specific about acknowledging a situation in your life:
I am grateful for a wonderful work environment because I get respect and appreciation for the hard work I do.

Or you might acknowledge having something: *I am grateful for having an apartment with lots of light because it keeps me in a good mood and keeps my plants happy, too.*

Remember, when you wake up in the morning and stop to identify positive things in your life, big or small, your mind will begin the day by processing the world differently. If you keep making your list in the same notebook or document on your computer, you'll see all the good in your life adding up.

Copy the following sentences and finish three each day.
Try this for one month and see what happens.

I am grateful for

Because

continued on page 156

CULTIVATING GRATITUDE: THINGS TO BE GRATEFUL FOR

I am grateful for

Because

I am grateful for

Because

The next exercise asks you to identify people that you're fortunate to have in your life.

- Is there someone that loves you, respects you and supports you?
- Is there someone who appreciates your uniqueness and individuality?
- Is there someone who acknowledges your hard work?
- Is there someone who simply makes you laugh and helps you enjoy the lighter side of life?
- Is there someone who is going through a difficult time and their experience has enriched yours?
- Is there someone whose company you don't always enjoy, but when you are around them, you learn a lot about yourself?

Give examples about why you're grateful that you have them in your life or that they came to you for a reason. Be specific. Start with people who are in your life now—family, friends, teachers, mentors, coworkers and so on. If you can't think of anyone in the present, think of people from the past—a teacher who made a difference in your life, someone who motivated you to choose a career or hobby you like, a childhood friend who moved away but taught you a skill you still use. Write down why each is special to you. How did the person affect your life in a good way?

Being specific about what you appreciate in the people you choose will ultimately expand your experience of them. In addition, if you're aware of precisely how people make you feel good, you may naturally end up giving some of that back. And giving back can undoubtedly inspire even more abundance because you may even feel grateful for the opportunity to make someone else feel happy. Studies show that the happiest people are the ones who help others.

EXERCISE 2
CULTIVATING GRATITUDE: PEOPLE YOU'RE GRATEFUL FOR

Write down three people that you're fortunate to have in your life.

As with the last exercise, do this for one month. For example, you might be specific about acknowledging why someone is so important to you:

I am fortunate to have my friend Jane in my life because she is a good listener, she treats me with love and respect, and she is there for me whenever I need her.

Or, you might acknowledge yourself:

I am fortunate to have myself because I have persevered this long, despite the fear and anxiety of a potential double-dip recession.

continued on page 158

CULTIVATING GRATITUDE: PEOPLE YOU'RE GRATEFUL FOR

Copy the beginning of the sentences below and finish them each day. Try this for one month and see what happens.

I am fortunate to have (person) in my life

Because he/she

I am fortunate to have (person) in my life

Because he/she

I am fortunate to have (person) in my life

Because he/she

The next exercise asks you to identify three factors that you believe make your life worth living. Try to give examples about why these are so important to you. Be specific. This third step is a personal values exercise, which unlike the other two steps, may or may not be goal oriented. Think about three values in your life that you could not live without. They must be so important that if someone tried to take them away, you'd fight for them. These values can't be material things like your money, car, house, iPod or even the indispensably assumed smartphone.

Again, the key is to be specific. For whatever values you come up with, try to identify exactly why they matter to you. For example, you might discover that your life is worth living because you like to help others, because you enjoy the work that you do in your career, because you enjoy being creative and expressing yourself by doing projects related to art or because you want to love and support your children and/or grandchildren.

EXERCISE 3
CULTIVATING GRATITUDE: WHAT YOU VALUE IN YOUR LIFE

Write down three reasons that you believe make your life worth living.

Try to give examples about why these are so important to you. Find anything in your life that you value and write it down. Other values could be that you are dedicated to personal growth and want to evolve more as a person. You may want to deepen your relationship with God or whatever your spiritual higher power is. Or you may find your life worth living because you want to overcome personal issues and obstacles that in the past have impaired your ability to be happy.

Copy the beginning of the sentences below and finish them each day. Try this for one month and see what happens.

My life is worth living

Because

My life is worth living

Because

continued on page 160

CULTIVATING GRATITUDE: WHAT YOU VALUE IN YOUR LIFE

My life is worth living

Because

Remember, acceptance also means being happy with what you have. But without taking inventory about all that's good in your life—big and small—you won't be able to appreciate it. These exercises work, especially when you practice them on a daily basis or as often as you can. They worked for me in the past, and to this day, they continue to help me to accept myself and manage my anxiety.

Chapter 10

REDUCING ANXIETY IN
INTIMATE RELATIONSHIPS

"Love is not a feeling; it's an ability."

—Peter Hedges

In the same way that intimate relationships can be joyful and rewarding, they can also be difficult to manage and sometimes even agonizing. The anxiety that can be caused by being in a relationship can also skew the mind into the negative thinking patterns that I've discussed in previous chapters. Close relationships tend to naturally feed into your vulnerabilities. The possibility of losing a relationship can create negative emotional responses and stress. Relationships can also trigger feelings that come from memories of your family interactions or relationships in the past that are painful or unresolved.

The risk you take when you fall in love and become emotionally dependent on someone else can be frightening. And you risk a lot if you place your hopes and dreams on another person. Sometimes, when there's discord between loved ones and you're worried that your most important relationship is on shaky ground, it's common to express your angst by getting angry. When this happens, the *fight* component of the fight-or-flight-or-freeze instinct kicks in because you may feel threatened that the relationship could end and you'll be all alone. So to reduce anxiety in intimate relationships, you must first look for any anger you may have and identify how it hurts you.

UNDERSTANDING ANGER AS A SECONDARY EMOTION

Anger is considered by many to be a secondary emotion since it may be triggered by key, primary emotions that are hidden away in the subtext of expressing your irritation. Anger is often merely the tip of the iceberg, with only a small portion of it on the conscious surface and a much larger, unseen percentage submerged at an unconscious level. When partners fight, they're not always aware of what they're really fighting about. Sometimes, beneath the surface of the familiar emotion of anger are very intense and often unspoken feelings of fear, hurt or both.

Getting beneath the surface of that emotional iceberg and becoming aware of the underlying feelings that we may have is perhaps what separates us from animals. You're not a walking impulse, incapable of thinking before you act. Therefore, you can rise above any habits you have of reacting instinctually and tap into the unconscious parts of your psyche that sometimes operate automatically. When your responses and behaviors are driven by anger, your emotions control you, which is likely to stoke the hurt or fear that you have below the surface. That increases anxiety.

Fear and hurt are by no means the definitive foundations of anger, although they greatly contribute to it. You may feel too uncomfortable or ashamed to express the real cause of your anger in the midst of a heated disagreement. Those unsaid thoughts can fester after a fight. In addition, sometimes one or both partners feel that expressing the fear or hurt might overexpose them and make them appear weak. In the passion of a highly charged quarrel, the ego comes in and pride takes over. Consequently, it's typical to turn to what's single-mindedly believed to be the most stoic and protective of emotions—anger. That makes you lash out instead of communicating how hurt you feel or what scares you. It happens to all of us, but that doesn't make it good for you.

In the heat of the moment it can seem better to give your anger the reins and let it speak for you. For example, in the midst of an intense argument you might shift into anger because you're afraid that the aggressive way you're being treated won't stop and things could escalate and get even worse. It can make you feel anxious, inadequate to change things, and subsequently, you can feel trapped. Your reaction may then be to make a negative statement instead of revealing how scared you are. What's important to understand is that, ironically, not withholding what you feel and sharing your hurt and fear with your romantic partner could actually increase the chances of repairing your relationship in the long run.

KELLEY AND ALAN'S STORY

Kelley and Alan came to me for couple's therapy, complaining of marital discord. While they both struggled with issues, Kelley began by sharing that she walked on eggshells around Alan most of the time without knowing why. After a few sessions, she had an epiphany—she was afraid to express negative emotions with him, especially anger. She felt embarrassed to admit it, but she feared retribution or even abandonment if she conveyed irritation or annoyance about anything. Kelley's mother had been a "grudger," who passive-aggressively punished her as a little girl by consistently ignoring her and acting indifferently whenever Kelley displayed any signs of discontent. She grew scared of her mother's withdrawal and over time blamed herself for temporarily losing her mother's love. She always worried that one day her mother would walk out and never return.

Kelley feared the same relationship dynamic with Alan, although he was nothing like her mother. During treatment Kelley recognized that withholding her emotions with Alan caused her and the relationship damage. I helped her get past her trepidation and to practice full disclosure of her feelings more with Alan, especially her anger. She began to feel less trapped by her old belief that anger was an inappropriate emotion, and it helped her get previously suppressed feelings off her chest. It also helped her because her fear that the relationship was always at stake ended. Consequently, their relationship improved drastically and they both reported to me that as a result of Kelley's openness, the marriage was strengthened and renewed.

It's also common to shift into anger if you're terribly hurt by what your partner said or did if his or her behavior or words remind you of being mistreated in past relationships or even by family members years ago. This is especially the case if you were abused, mentally or physically. Bad memories can cause you to pick up on things your partner does and interpret them in ways that are way beyond their intention. One word can remind you of someone who did terrible things to you. One action can make you scared that the hurt you experienced will be repeated. Those memories can push angry buttons that trigger you to protect yourself.

You may believe that an expression of anger means you're standing up for yourself, so you don't share how truly wounded you feel. But if you keep these feelings bottled up and never express them, anxiety gets activated because they're still there, underneath the surface of your feelings. Suppressing how you really feel leads to feeling powerless over what's going on, and you can easily start to feel overwhelmed. As a result, your conflict-resolution skills become two-dimensional: you're either silent about the issue, or otherwise you explode in anger and your true feelings aren't addressed.

For most people, expressing anger openly has, at times, brought results that got their needs met. As discussed in Chapter 8, you may have let anger drive you to use an aggressive communication style to protect yourself and make sure you're taken seriously. But typically, inappropriate words or angry behavior don't foster intimacy in a relationship and can establish a pattern of relating with loved ones in unhealthy ways that have a long-term negative effect. Therefore, it's important to learn how to anticipate situations where you let anger communicate for you and prevent them by focusing on the anxiety that anger-based communication produces instead. You may not be able to stop the anger itself because you don't always have control over your feelings. But you can control what it makes you do. Remember, there are no wrong feelings and no wrong thoughts. There are only wrong actions.

I used to have a chronic fear that if I expressed any kind of anger, ever, I'd resemble my emotionally unstable father and over time become him, too. Yikes! My father was famous for his uncontrolled fits of rage that were embarrassing to watch and painful to experience when directed at me. He often humiliated me and my family in public by spewing unspeakable rage if he was upset with a waiter's poor service or if he was dissatisfied with a salesperson at a store or anyone that "wronged" him, period. As a result, I naively engaged in all-or-nothing thinking by *never* expressing anger at all to ensure I didn't transform into a sadder version of him. I never wanted to make a spectacle of myself like he did. The price I paid was suppressed emotions that increased my anxiety. Recognizing that fear connected to my relationship with my father helped me get past it, and in time I managed to give it up.

Anger that festers without resolution can fuel anxiety. Buddhist monk and peace activist Thich Nhat Hanh gives an example of how seeking punishment through anger can be self-destructive. He said that if someone sets your house on fire, the first thing you should do is try to put out the fire, not run after the person you suspect is the arsonist. If you run after that person, your house will burn down while you chase him or her. It's more logical to put your immediate energy into dousing the fire. Getting angry as a result of other emotions keeps you from putting out the fire of your anxiety. Instead, your anger stokes it. If you can embrace the fact that your well-being is most important, you can motivate yourself to begin to speak up in an honest, noncombative tone to your partner about what bothers you. When you clear the air, you also clear the reasons for being angry.

JOSH'S STORY

Josh came to me because he was frustrated by his inability to make a relationship work. He had been burned years ago by a woman he deeply loved and was determined not to let someone hurt him again. He had many girlfriends since Courtney, but each time things got serious, he'd get nervous and notice things that reminded him of her. His inappropriate accusations and anger made them all leave. He really liked Amy and didn't want to lose her. But he got angry with her a lot and she was getting tired of it, often saying that she didn't want to pay for the bad behavior of women before her.

He asked with total conviction, "Don't I owe it to myself to be careful to not let a woman hurt me again?" I explained the difference between looking for unacceptable behavior and letting the anger he had with Courtney drive his perception of other people's behavior. And, the anger kept him from openly acknowledging what was going on with the women he dated and therefore prevented them from being able to work it out. His anger was on autopilot, to protect him. But his reactions just hurt him more when he couldn't make any relationship work.

He was scared of losing Amy, which also triggered anger since it made him feel vulnerable. Yet he knew he was driving her away. He shared that one day Amy left a message that she'd be home by

7:00 if he wanted to call her. He'd just landed a new client and was excited to share the news with her. But when he called at 7:00, Amy wasn't home and didn't return his call until after 8:00. By then he was fuming. Courtney had never kept her word. So he yelled at Amy, without even hearing why she had been delayed. He later apologized but didn't explain why it bothered him so much. She called him irrational and didn't think she could stay with him.

Josh knew that Amy was good for him and didn't want to lose her. He asked me for help. I suggested he first accept that Amy wasn't Courtney and he needed to stop looking for signs that she was. The most important thing, as hard as it might be, was to have an honest talk with Amy and explain why he behaved as he did. Doing that terrified him, but the thought of losing her felt worse. Josh knew on a rational level that Amy wasn't like Courtney, and he kept reminding himself of that. He invited her out to a quiet dinner and told her she meant enough to him that he was willing to open up, which scared him. He then shared his story. Amy was touched by his honesty and agreed to work with him to help him stop his overreactions.

Josh was relieved that she knew the truth. They came up with things that Amy should say if Josh went into his old patterns to remind him she wasn't Courtney. She was also more sensitive to how he reacted to things not going as expected and communicated more if she had to do something different or would be late. It took time, but with patience, Josh let go of his anger about how Courtney hurt him so that he could let Amy's love in. They've been living together for two years and plan to get married soon. Josh still has residual hurt feelings, but they continue to subside and he rarely feels anxious about his relationship. He's happy to no longer be the guy who can't keep a girlfriend.

CHANGING FIXED BELIEFS ABOUT RELATIONSHIPS

One of the goals for couples interested in reducing tension and achieving longevity together is to cooperatively begin the process of taking the myths—the principles you've lived by that aren't based in truth—out of any fixed thinking from your personal belief system. It's time to let go of any consensus reality you were taught about how a marriage or a relationship "should" be or look like. For example, a common relationship myth is that if there's deep and committed love for each other, then all problems will naturally go away and everything will work itself out. Not true. Many people believe that a healthy relationship means no arguing or fighting and a couple has to agree all the time. While you don't need big fights, disagreements are actually healthy for a relationship. When a couple brags that they never have an argument and agree with everything, I wonder what feelings they're keeping quiet that will eventually emerge.

Another relationship myth is that being in love means that you must take responsibility for making each other happy. This is false, too. The truth is, while love is a powerful force, it's not always very practical. You don't have the ability to make someone else happy, although you can add some happiness to his or her life. Yet you do have the ability to make someone else miserable. So to reduce anxiety and increase your chances for long-term success with your partner, there are six tips that I strongly urge my patients to adhere to. They will help you avoid making each other miserable.

These tips are designed to decrease tension by asking both partners to heighten their awareness of each other, to reexamine their personal beliefs about what a relationship "should" look like and to take personal responsibility for their own happiness. Of course, most couples want their relationships to succeed. Many desperately want them to work out, but the only thing that will ensure this is if both agree to change their thinking and subsequently the way they behave with each other. Without a behavior change, there's no growth. And no growth means the relationship will falter over time. Accept that love is not just a feeling—it's an ability.

It's important to understand that the following relationship tips are only effective if *both* partners are dedicated to the relationship enough to be willing to change and grow together. One person can't do the work of both partners to have a solid, healthy relationship. There must be honesty, fairness and goodwill from both sides.

1. **Both partners must commit to the practice of lowering their expectations.** Often when you fall in love, you think the person you're in love with is perfect. Everyone is on their best behavior at first. But that image can slowly unravel over time, since no one is perfect. That can cause unnecessary static if you feel like the person is changing. In a good relationship, each person must put aside wanting their partner to be perfect in all areas and begin the process of accepting him or her for everything they are and everything they are not. Lowering your expectations also means lowering your own expectations of yourself, too. You can't expect to be the ideal mate either. Needing to be perfect and anything in the vicinity of striving to be so has no place in any kind of human relationship, whether it's you or your partner. We're simply not built that way.

   We are in fact, exquisitely flawed. It may sound like an obvious requirement for an intimate relationship, but it actually takes a great deal of effort to break through previously held beliefs about how the other person should be or how you should be. Both sexes create fantasies about who they want romance with. These fantasy beliefs are ingrained in your psyche, but since you weren't born thinking this way, they can also be unlearned.

Lowering your expectations doesn't mean lowering your standards about the kind of qualities you want in a romantic partner, like honesty, integrity, intelligence and loyalty. Nor does it mean settling for someone who you're not compatible with. It also doesn't mean compromising your personal values and beliefs about how you want to be treated or even how you want to be loved. It simply means having realistic expectations of each other that are reasonable and practical. It means embracing each other's imperfections as one of the many wonderful variances that make us human.

2. **Each partner must begin to let go of the need to be "right."** The old question still holds true: "Do you want to be right? Or do you want a relationship?" You can't have both, period. Needing to be *right* assumes a one-up position of superiority and power that causes people to feel talked down to or even attacked. When someone needs to always be right, he or she is being inflexible about the natural path of personal growth and transformation everyone must go through. Since no growth means no change, no change may mean no relationship. If one partner won't bend and the relationship becomes inflexible, it can tend to break easily. It's also important to remember that wanting to be right represents a dualistic mindset because you live by the extremes of being right or wrong.

If thinking with a mindset that looks through a lens of seeing everything as being right or wrong causes you anxiety, it will most likely cause the same kind of anxiety in your relationship. Giving up the need to be right means each partner must let go of trying to control the other and trying to get your partner to continually see things your way. It's also essential to accept that you're very different people, and that's okay. If you and your partner can let go of any need to always be right, you'll both benefit from celebrating the uniqueness you each have and not make each other wrong for it. This requires working together to tolerate each other's conflicting frames of reference, especially on touchy subjects that have caused discord in the past.

Regardless of how deeply emotional an issue is between you and your partner, removing the unbending stubbornness of wanting to be right all the time gives the relationship a better chance to survive. Something I've heard from many individuals that I've treated is that deep down inside, they truly wished their partners were more like them. And they believed that if the partner was more like them, they'd have more harmony in their relationship and problems would somehow disappear. This is also a myth.

3. **Both partners must commit to the practice of acknowledging their own part or contribution to the problem.** Each partner must first identify and then hold himself or herself accountable for how they negatively contribute to the discord in the relationship.

This involves taking an honest, close look at what you yourself are responsible for that only you can change. Otherwise, the relationship becomes one of blaming each other (he said/she said) and expecting the other person to change. A healthy relationship has close to a 50-50 split on taking responsibility for the problems between you.

In order for any relationship to work, and in order for change to be able to take place as it goes along, the relationship must start as the Latin expression *tabula rasa* states: with a clean slate. To have that clean slate and keep it fairly clean, both partners need to commit to working on making changes in his or her behavior, without focusing on the other. This means that each partner is only responsible for cleaning his or her own side of the street and becoming accountable for himself or herself.

You don't have the power to fix or change your partner, but you absolutely have power over yourself. How much power is your choice. And that power can make changes in you that may positively influence your partner. My patients argue that not being able to change their partners makes them feel powerless over how their relationship goes. I assure them that it's actually quite empowering. As a therapist, I'm solely a facilitator, not a fixer. This means I try to guide patients, like I'm guiding you, to find ways to create conditions in their lives that will allow them to have healing that leads to changing the dynamics in their relationships for the better. By working on your own problems, you help smooth the progress of

your relationship—you're not trying to fix it. It's a subtle
but important distinction to make. When you both commit
to finding answers in yourselves, you have a much better
chance to have a less stressful relationship.

4. **Each partner must commit to the practice of monitoring
his/her own personal blind spots.** You must be willing to
learn to recognize what pushes your buttons about your
partner—what is it that makes you irrational and unable to
be realistic, such as when, say, your spouse leaves his or her
dirty socks on the floor for the fifth day in a row. Not being
aware of what I call your blind spots (like pet peeves) in
a relationship leads to unnecessary and unwanted reactions
when they occur. For example, one partner may demand
extreme neatness and cleanliness in the house. If his or her
spouse pays little attention to cleaning up, it can trigger
a very knee-jerk, rash response and easily irritate that person.

That sensitivity to anything that's not perfectly clean
and in its place can result in overreacting to what could be
considered a blind spot. This person gets incredibly flustered
when something isn't "right." Having a heightened awareness
to those personal hot buttons can preempt an argument. Over
time, both partners can then compassionately acknowledge
each other's blind spots or hot buttons and know when to
pull back and respect the other's boundary. Since people
need to be responsible for themselves, each person needs to
monitor his or her own hot buttons. It's not up to you to keep
an eye on your partner's and try to do damage control.

In order to have a good chance for developing a healthy
relationship that isn't a big source of anxiety, you have to be

accountable for yourself. That means taking responsibility for your choices and actions within the relationship. Relationships work much better and less tension gets generated between partners when each of you takes on that mission and self-monitors in the same way you'd identify your negative contribution to the problems as stated in the third tip. When you're both on board with handling your hot buttons this way, there's no pressure on each other. A good mantra to use together is, you focus on what you want to change, and I will focus on me.

5. **Each partner must commit to the practice of using assertive communication.** You'll both have to change the way you communicate and learn to give up communicating aggressively or passively if you want to keep things between you on a positive path. Aggressive communication usually starts a heated argument (or worse) and causes bad feelings. Passive communication means you hold what bothers you inside, but it will still guide how you treat your partner because it also stirs up anger. There is also the gray area communication style called being passive-aggressive, which is when individuals relate anger or negative feelings in an indirect way. It can be communicated through sarcasm, humor, resigned indifference and sometimes by simply ignoring the other person. Either way, it's a kind of sneaky way to be heard, thinking it may avert an all-out heated argument. But like aggressive and passive communication, it doesn't help the relationship in the long run.

I've always considered healthy communication as the oxygen that keeps relationships breathing and thriving.

The quality of that communication is very important—it's the backbone of a solid partnership.

Assertive communication involves using nonjudgmental "I" statements to relate to each other, as discussed in Chapter 8. "I" statements are key because they help the speaker take full responsibility for his or her feelings and thoughts without the landmine of using "you," which could be seen as blaming and putting responsibility for what causes static in the relationship on the other person. Blaming or constantly expecting to be blamed for something with aggressive communication can cause fear and anxiety, which creates stress in the relationship.

Again, if your partner is a neat freak and gets upset if anything is less than perfect, you might feel underlying stress whenever you're home. You may worry that you left something out of its place or didn't tidy up enough. If you know your partner gets angry at those things, you can become obsessed, worrying that you'll be verbally attacked for it. That's why it's important to use assertive communication to tell your partner how you feel and that you don't want to continue feeling all this stress. Ask for a way to compromise, since being overly neat isn't your way. If you don't point the blame at him or her, you might get through.

Another example is having a partner who is a stickler for punctuality and hates to wait when anyone, including you, is late. You may have a different concept of time and commitment that does not place so much emphasis on exactness. Instead of walking on eggshells around this issue and feeling afraid to be late, assertive communication may help balance your partner's rigidity on this subject. If you clearly, without being critical, explain why it bothers you so much and why it's stressful to deal with, that understanding may motivate your partner to try to lighten up a bit.

In the short run, taking responsibility for whatever your part is in an issue while using assertive communication helps partners begin to respond to each other instead of react. Most couples don't know how to talk to each other without triggering an argument. I've helped many couples over the years to improve their relationships by simply altering their communication style to being more assertive and less aggressive. The anxiety and tension levels that are immediately reduced by this fundamental modification in communication is remarkable. If there's goodwill between partners and both want to change, this aspect of improving a relationship usually works.

6. **Each partner must commit to cultivating friends, activities and personal interests outside of the relationship.** This tip is one that I usually get a great deal of resistance to, but it's vital for reducing anxiety in relationships. This prevents putting too much pressure on each other for fun and stimulation. Often partners buy into old beliefs that they should be together all the time, rely on each other for love as well as for friendship, and share the same interests and hobbies. They're scared that if they don't, something is wrong. This is simply another silly myth that does more harm than good.

Each partner should try to cultivate interests outside of the relationship in order to make being together fun again.

This creates more breathing room in the relationship and prevents being dependent on each other for fun. Spending time together should not feel like an obligation. I strongly urge my patients to create conditions in their relationship so that wanting to be together is something they can look forward to. When you're away from your partner enough to miss him or her every now and then, it can kindle the excitement of seeing each other again. It's healthy to plan date nights on weekends like you did when you first met. This keeps the relationship fresh and stimulating.

As discussed earlier, the personal spaces that exist in between two people who love each other are critical to the longevity of a relationship. Many couples tend to mistake this for cultivating separate lives that alienate their connection. That's far from the truth! When your mind is balanced, it doesn't default into a dualistic extreme of desiring too much space that will distance you both. In a sense, having autonomy within your relationship is like finding the Goldilocks Zone (discussed in Chapter 4) of your relationship—where both of you can thrive by being true to your individuality and also commit to loving each other.

ANGRY MIND VS. BALANCED MIND

Anger itself can create so much irrationality in one's thinking that couples are ill equipped to manage it without the awareness I've presented. When anger isn't recognized as a byproduct of the hurt and fear you have, then each of you might confuse the anxiety it produces with needing to assert yourself negatively. That can make you want to control your partner, create a strong need to be right, develop unrealistic expectations and communicate aggressively.

Anger produces a fervent response for holding your ground and fighting to the bitter end. The raw emotion experienced creates a mindset that automatically triggers the fight response. Whereas a balanced mind helps you find healthy ways to work with your partner, an angry mind is likely to make you feel defensive and shift into a results orientation that pushes you to develop control strategies. When you're in this frame of mind in a relationship, you might:

- Seek perfection and develop high and unrealistic expectations for your partner.
- Need to be right all the time and think in black-and-white terms.
- Try to control and push your partner to be more like you.
- Blame your partner and not take responsibility for the problems in your life.
- Use aggressive or passive communication (ignoring or being aloof).

A balanced mind lets go of a need for control. Anger then gradually decreases as the anxiety is reduced. A balanced mind leads to using process orientation and helps you give up any need to control your partner. When you're in this frame of mind in a relationship, you:

- Let go of a need for perfection and develop realistic and practical expectations.
- Balance any need to be right and look for the gray areas that aren't dualistic.

- Give up trying to control the other person and focus on what you do have control over.
- Stop the blame game and take responsibility for problems you've contributed to.
- Use respectful and honest assertive communication.

Chapter 11

ANXIETY-REDUCTION
MAINTENANCE

How to Keep It Going

*"Life is ten percent what you experience
and ninety percent how you respond to it."*

—Dorothy M. Neddermeyer

Imagine that you're the director of a film about your life and, until now, your field of vision through the camera lens has been limited by your free-floating anxiety. Also imagine that due to the types of imbalanced thinking styles discussed in previous chapters, your film has been restricted to a series of narrow, close-up shots that have deprived you of seeing past what's directly in front of you and, more important, seeing what's around you. Opening up your mind and altering your thought process can be like pulling the camera back to a wide-angle view of your life and embracing the panorama—the big picture.

BECOMING THE DIRECTOR OF YOUR LIFE

As the director of your film, knowing how to access the big picture when you need to helps to reduce anxiety by allowing you to edit the story and cut down anxious thoughts to size and decrease their proportions. The close-up vision doesn't allow for clear, rational thoughts, so you may tend to react as if you're trapped. Once that happens, anxiety creeps in, and fear can come soon after. Zooming out and viewing the big picture provides alternative and open-ended possibilities that help you see beyond the limiting walls of your negative thinking.

In this chapter, I'll prescribe daily maintenance exercises to assist in the process of accessing the bigger picture of your life when you start to get anxious thoughts or experience emotional reactivity. Their purpose is to help pull back the camera from the close-up on your mind and instead restructure your automatic responses to stressful situations. By completing these exercises on a daily basis and maintaining a steady routine of checking in with yourself, your mind can shift from being a victim of your thoughts to being an observer of them. This shifts you from anxious reactions to rational reflection.

The following three writing exercises are based in traditional cognitive behavioral therapy techniques developed by Aaron T. Beck and rational emotive therapy techniques developed by Albert Ellis. Their groundbreaking techniques have been widely used for a variety of conditions and have become therapeutic standards for many clinicians. These types of exercises have proven to be very effective if used consistently over a significant period of time. I've put my own spin on them.

OBSERVE AND IDENTIFY EXERCISE

Begin the first exercise by identifying your negative thoughts and categorizing them according to three of the concepts I've discussed throughout this book:

- Dualistic mind
- Consensus reality
- Illusion of control

See if you can distinguish one thinking style from another. When you feel anxious or stressed about anything, big or small, try to shift into observer mode. Pay attention to it, identify what it is and log it in. Over time you'll sharpen your thought-identification skills and begin to grasp that your beliefs, which create the thoughts, are part of the old ones you made up years ago. So when you observe and identify any negative thoughts that you can recognize as falling under those categories, you will then try your best to dismantle those beliefs and take away their power.

MAKE YOUR OWN WORKSHEETS

Use the following worksheet by filling in the boxes assigned to the three negative thinking styles. Identify which of the three styles applies to your situation and state the facts about the stimulus that triggered your anxiety. Then define the negative thought or the negative interpretation that you have attached to the situation. Observing and identifying your negative thinking styles in this exercise will prepare you for the Replacement Thoughts exercise that follows the worksheet.

In the worksheet you'll see examples of three negative thoughts attached to three situations under all three categories. Use the examples as a guide to help you learn how to observe your own negative ones. Copy the worksheet into a notebook or in a document on your computer and use it for your own thoughts every day.

DAILY DIARY OF THOUGHTS:
EXERCISE #1
OBSERVE AND IDENTIFY

Observe	Notice your thinking this week
Dualistic Mind—Thinking in Extremes	**Observe:** Did I use extreme thinking today (such as *good* or *bad*, *right* or *wrong*, *strong* or *weak*, *smart* or *stupid*)? **Situation:** Due to an unexpected work emergency, I was late picking up my children from school. **Negative thought:** I feel like a bad parent when I have to make my children wait for me. A good parent would *not* be late.
Consensus Reality—Clinging to Fixed Thinking	**Observe:** Did I engage in using absolute words such as *never*, *always*, *should*, *shouldn't*, *everyone*, *no one*? **Situation:** I attended a wedding and a baby shower for two close friends in the same month. **Negative thought:** Every one of my girlfriends is married and having kids. At thirty-two, I *should* be married with a family by now. Now I am too old and I will *never* meet Mr. Right. *All* men want younger women. I will always be alone. I *shouldn't* have listened to my parents and gone to law school.

continued on page 186

OBSERVE AND IDENTIFY

Observe	Notice your thinking this week
Illusion of Control—Seeking Control in Results Orientation	**Observe:** Did I try to seek results by trying to be perfect, being a people pleaser or worrying about the future? **Situation:** My boss looked upset this morning when I arrived at work. **Negative thought:** He is upset with me for not performing well on the job. I am going to get a poor review and I will not get the promotion or I will get fired. My life is over!

Date/Time:

Your Situation:

Your Negative Thought:

Your Response:

REPLACEMENT THOUGHTS EXERCISE

After you've practiced observing your negative thoughts and labeling them according to the three categories, in the next exercise you will respond to the thoughts with Replacement Thoughts via the Five-Minute Rule. This two-step exercise is training your mind to not leave negative thoughts unchallenged, since these kinds of unchallenged thoughts tend to snowball throughout the day and escalate your anxiety. For this exercise, every negative thought is seen as an opportunity to reframe it for use as a tool instead of having it be a hindrance. It trains you not to run from your anxious thoughts, but to take advantage of having them.

Remind yourself, *I can be reflective, not reactive.* Automatic negative responses via the three categories described previously cause reactivity and escalate stress levels. But grounded statements that are contemplative in nature slow down the reactivity, reduce emotional distress, and cause you to reflect on the situation, which ultimately reduces stress levels.

Integrating Replacement Thoughts into your mind is like having to restart your computer in order for the new software you just installed to take effect. A restart does not mean shutting down your computer. It simply means you're adding a new program. And with the new Replacement Thoughts you've cultivated, you'll not only see the big picture mentioned earlier, but you'll also begin to establish new ways of thinking that can ultimately force you to create your own reality.

As with Exercise 1 in this chapter, any anxiety will be your cue. When you feel it coming on, break out this worksheet and do step one—identify the negative thought you have based on the situation. Then do step two—take five minutes to reflect on the situation and replace that thought with a grounded, more realistic thought that does not include:

- All-or-nothing words like *good* or *bad, right* or *wrong, strong* or *weak* (dualistic mind)
- Absolutes like *should, never* and *always* (consensus reality)
- Seeking perfection, people pleasing and immediate results (illusion of control)

Remember that a Replacement Thought is a challenge to the old ways of thinking that you've clung to for many years. It may not come easy at first, but over time, if you formulate grounded statements that don't involve the kind of language used in the three categories described previously, you'll feel less anxious. Be mindful about not allowing any thoughts that speak to a need for perfection or illusion of control to creep in while you do the exercise. Look for progress, not perfection. For help with formulating Replacement Thoughts, review the examples given in Chapters 3, 4 and 5. These are very generic ones, and your own thoughts need to be expanded to include explicit details of your unique situation.

In addition, it may be helpful to review Chapter 7, since it includes Replacement Thought examples that go into deeper and more detailed feelings that can help you formulate your own. In Exercise 2, you'll see an example of three negative thoughts attached to three situations under all three categories with an example of a Replacement Thought. Make your own grid in a notebook or on your computer.

DAILY DIARY OF THOUGHTS: EXERCISE #2
REPLACEMENT THOUGHTS—FIVE-MINUTE RULE

Identified Negative Thinking Style #1	**Dualistic Mind—Thinking in Extremes:** Am I using extreme thinking today (such as *good* or *bad*, *right* or *wrong*, *strong* or *weak*, *smart* or *stupid*)?
Activating Event: What happened today that got me so upset?	**Situation:** Due to an unexpected work emergency, I was late picking up my children from school. **Negative thought:** I feel like a *bad* parent when I have to make my children wait for me. A *good* parent would not be late.
Replacement Thought—Five-Minute Rule: Take time to step back and respond reflectively instead of reactively.	**Response:** Wait! Stop! There I go again. Just because I am late once in a while does not make me all bad. There are many other ways I am a loving and conscientious mother. **For today, I'll accept that being a mother means that my peace of mind rests in the gray areas of my life. I'm not good or bad: I am okay.**
Identified Negative Thinking Style #2	**Consensus Reality—Clinging to Fixed Thinking:** Am I using absolute words such as *never, always, should, shouldn't, everyone, no one*?

continued on page 190

REPLACEMENT THOUGHTS—FIVE-MINUTE RULE

Activating Event:	**Situation:** I attended a wedding and a baby shower for two close friends in the same month.
	Negative thought: *Every one* of my girlfriends is married and having kids. At thirty-two, I *should* be married with a family by now. Now I am too old and I will *never* meet Mr. Right. *All* men want younger women. I will *always* be alone. I *shouldn't* have listened to my parents and gone to law school.
Replacement Thought—Five-Minute Rule:	**Response:** Slow down! Who says that thirty-two is too old to get married and have kids? And where is it written that one should be married by thirty-two anyway? These days women are getting married and having kids in their 40's. Saying I will *always* be alone is an "absolute" statement.
	Plus, choosing to become a lawyer has been my lifelong dream. There's nothing wrong with being ambitious.
Identified Negative Thinking Style #3	**Illusion of Control—Seeking Control in Results Orientation:** Am I seeking results by trying to be perfect, being a people pleaser or worrying about the future?

Activating Event:	**Situation:** My boss looked upset this morning when I arrived at work.
	Negative thought: He is upset with me for not performing well on the job. I am going to get a poor review and I will not get the promotion or I will get fired. My life is over!
Replacement Thought—Five-Minute Rule:	**Response:** Hold on a minute! I can't be perfect on the job but I know that I've done my best. If I don't get the promotion, it doesn't mean my life is over. My self-worth as a person is not dependent on my job. I am good at many other things.
	Plus, how do I know for sure my boss is mad at me? I have no control over my boss's moods. I don't have that kind of supremacy over anyone.

Date/Time:

Your Situation:

Your Negative Thought:

Your Response:

BREAKING DOWN YOUR THOUGHTS

Now that your negative thought identification and Replacement Thought skills are getting sharper and you've practiced with the first two exercises, it's time to take the next step by using a handy worksheet that will deepen your ability to reduce anxiety even more. This exercise is slightly different than the last one and more accessible so you can take it with you and fill it out any time of the day. This exercise is the go-to, nuts-and-bolts worksheet that my patients have used most often. It's easy and quick to fill out and literally breaks down a thought into five easy-to-understand stages. It's the most efficient way to help you de-escalate your symptoms in the moment by grounding you in a here-and-now task.

I've discovered that in the midst of breaking down thoughts in this fashion, the process itself can disarm a negative thought and strip it of its power and credibility. Consequently, as time passes, it can help you stop the racing thoughts and calm you down more rapidly. Instead of allowing yourself to be vulnerable to your spinning anxious thoughts by white-knuckling it on your own and hoping the angst will go away, the act of filling out this worksheet will empower you. This can help you get so familiar with breaking down your thoughts this way that you begin to do it automatically in your head without needing the worksheet in front of you.

This exercise adds two new features to the mix—assessment and evaluation. The assessment stage of the thought breakdown asks you to rate your anxiety level on a scale of 1 to 10 (1 being very low anxiety and 10 being very high). This gives you a point of reference to be able to see your anxiety decrease every time you use the worksheet. The evaluation stage also asks you to rate your anxiety level after you've cultivated a new Replacement Thought. As you complete the worksheet, you'll see the evaluation anxiety rate levels decrease from the assessment levels. Break down your thoughts in a five-part worksheet using the acronym SNARE.

Situation
Negative thought
Assessment
Replacement thought
Evaluation

Situation: Write down the actual situation or event that triggered your anxiety, only stating the facts about what happened. It doesn't involve your personal opinion or any interpretation or appraisal of the situation—just the physical truth. Keep it as basic as possible, such as *Stuck in freeway traffic and running late for my appointment with a client.*

Negative thought: Write down the immediate automatic negative thought that pops into your mind as a result of the aforementioned situation. As in previous exercises, the negative thought will have aspects of one or all three of the negative thinking styles discussed in this chapter. The negative thought is, of course, your interpretation of the situation and it's often your irrational or distorted view of the event or trigger. The negative thought stage is the lynchpin to your anxiety rising and falling. It's the focal point of your angst, not the situation.

When documenting your negative thought, try to push the interpretation as far as you can—be as irrational as possible and take your most fearful thoughts to the limit. Allow it to snowball until it starts to sound absurd. The more absurdity you can put into your thinking, the more you dismantle its power and credibility:

Oh, no, I'm always late. My clients won't take me seriously and I'll look irresponsible.

Then push it further and ask, what will happen next?

If my clients don't take me seriously and I look irresponsible, I'll be humiliated and lose my job.

Then you will push it further again and ask, what will happen next?

If I get fired from my job, my wife will leave me and I'll be all alone.

Next?

If my wife leaves me and I don't have a job, I'll lose everything. I'll be homeless and fade into obscurity and perhaps even die an early death.

Clearly, the last response is the most extreme, the most reactive and the most absurd. This would be the one to log in your worksheet. Most of the time when I ask my patients to push the envelope and snowball the negative thought until it starts to sound absurd as I just demonstrated, they resist because it's too scary to do. They're used to leaving negative thoughts unchallenged, which is a mistake. So I remind them that it's critical to snowball it because while you may not be conscious of it in the moment, the ultimate fear of ending up alone or dying an early death seems to often be what people fear the most. It's the common theme at the tail end of most extreme negative thoughts.

Leaving the thought unchallenged allows it to fester in your mind, and it remains hazardously untapped. I always say, "Get to the root and give it the boot." A turning point is when you start to see that one reason you get so anxious is that the perception you create with your thoughts can often outrageously jump from being late for an appointment to losing everything that's precious to you. When that becomes clear, your rational awareness can kick in and new thinking can start to unfold. This helps you see that, in reality, it's too big a leap, and balanced thinking also exists in between that unreasonable chasm of thinking. So the more irrational you get during this exercise, the better. And doing it over and over further helps dismantle the negative thought.

Assessment: Pause and use the Five-Minute Rule to reflect on the thought. Define what you feel and rate your anxiety on a scale of 1 to 10. Rating your anxiety provides a beneficial point of reference that helps you observe how your anxiety rises and falls simply by what thoughts you entertain.

For example, thinking you may lose everything that's precious to you because you're stuck on the freeway will most certainly produce at least an 8 or 9 on the anxiety scale. Reflecting or using the Five-Minute Rule may trigger you to recognize that if you're simply stuck in traffic and if your anxiety is high, this must mean more than just being scared of being late. That insight will be the springboard for the Replacement Thought, which is the next stage.

Replacement thought: Creating Replacement Thoughts is often the hardest part of the thought breakdown. But by now, you've had practice with formulating your own rational thoughts and begun the process of creating your own reality. Your Replacement Thought is the rational and big-picture response. A Replacement Thought challenges the negative thought belief that fuels your anxiety.

This can be the most difficult stage. You're likely to struggle, at least at first, to come up with rational thoughts on your own. One of the tricks you can use is to pretend that you're helping a good friend come up with some. Ask yourself, if a close friend came to me with the same situation and was having the same type of extreme negative thoughts, what would I say to him or her? More than likely, you'd find it easier to help your friend come up with kinder thoughts than you would for yourself. So let's say your friend goes through the example of being late because of traffic and tells you:

> *If my wife leaves me and I don't have a job, I'll lose everything. I'll be homeless and fade into obscurity and perhaps even die an early death.*

As a caring person, you'd probably respond with compassion and may offer some Replacement Thoughts that may actually come to you without trying hard to refute that thought. If you care about this person, you'd probably suggest a more rational thought because you'd want to help your friend feel better. So that same conscientiousness you possess must be used for helping yourself see a better picture. An example Replacement Thought might be:

> *Wait! Slow down. "Lose everything" is an absolute statement. That's my anxiety talking. Being late for an appointment does not necessarily mean there will be negative ramifications. I can call ahead from my cell phone and let them know there was unexpected traffic. Plus, being stuck in traffic that makes me late has very little to do with losing my wife or dying.*

Evaluation: This last stage is simply evaluating how you feel after coming up with and accepting the Replacement Thought. Now rate your anxiety on the 1 to 10 scale again. Then frame a conclusion about the entire breakdown of the thought itself.

- What do you think about the situation now?
- How do you view the situation when you look at the big picture?

For example, an evaluation entry might be:

> *My anxiety level is slightly lower at 7. And although I may get anxious again if I'm a little late, I can manage it better by identifying my negative thoughts. I'm realizing that I have a tendency to overreact and blow minor situations out of proportion. I need to use the Five-Minute Rule more often and pause before I jump to anxious conclusions.*

The following is the complete SNARE method exercise that includes the assessment and evaluation sections. Copy it into a notebook or in a document on your computer. Leave enough room to write down everything you come up with for every day. Use each line starting in the situation stage to log in three different situations throughout the day. When you write down yours, make entries for each stage that follows to correspond to each number. For example, the situation referred to in #1 will correspond to #1 all through the exercise, and the same with #2 and #3. The #1 in the exercise is a sample that you can use as a guideline for your own thoughts.

DAILY DIARY OF THOUGHTS:
EXERCISE #3
REPLACEMENT THOUGHTS—ASSESSMENT AND EVALUATION:
SNARE METHOD

Situation: An event or situation that triggers stress and anxiety. Note just the facts.

1. **Sample:** *I am stuck on the freeway, and I will be late for an appointment.*

2.

3.

Negative thought: Write your automatic negative thought about the situation. What is your personal interpretation of the event? Be as irrational as possible. Allow your negative thought to snowball until it sounds absurd.

1. **Sample:** *If it makes me late, I'll get fired and lose everything. I'll die an early death.*

2.

3.

Assessment: What are you feeling as a result of the negative thought? Use the Five-Minute Rule to reflect and assess. Rate the degree of anxiety on a scale of 1 to 10.

1. **Sample:** *I am scared and anxious. My anxiety is at a 9!*

2.

3.

Replacement thought: Enter a rational (big picture) response that challenges your negative thought. Pretend you were giving the same advice to a good friend. What would you say?

1. Sample: *Stop! Being late is sometimes part of life and could not be helped today. I am blowing this out of proportion. "Losing everything" is an irrational statement. I will call on my cell and let them know ahead of time. Beyond that, I have no control over this situation, period.*

2.

3.

Evaluation: What's your anxiety level on a scale of 1 to 10 after implementing a Replacement Thought? What's your conclusion? What do you think of your negative thought now?

1. Sample: *Okay, my anxiety is still high, but it has decreased to a 7. I am okay once I take five minutes to reflect. I am starting to see my patterns of reacting too fast.*

2.

3.

Now that you have tools to identify, break down and de-escalate your negative thoughts on your own, test it for several months. Fill out the worksheets as best you can, whenever you can. Pay attention to your anxiety rating scales. As time passes, are you scoring lower when anxiety is initially triggered? And once you create a Replacement Thought, does it decrease quicker? During those months, it's important to structure your time wisely, too. I recommend keeping a Daily Accountability Diary, which I'll describe later in this chapter.

My patient Rafael is a good example of someone who used this effectively. He used to white-knuckle his anxiety episodes by innocently hoping he could somehow will his mind to stop racing with stressful thoughts. He discovered that merely wishing and praying for the anxiety to go away on its own didn't work. These exercises helped because they forced him to act instead of being victimized by his negative thoughts. Writing down his negative thoughts helped disarm their power. By blueprinting his thought processes and seeing how simple yet complex they were, especially with the SNARE method, he grounded himself in the here and now and got a visual of how absurd his anxious thoughts really were. He often said that the most significant part of the exercise for him was that it forced him to cultivate Replacement Thoughts that he'd never have imagined on his own. It opened up his mind to new possibilities of thinking that surprised him. And usually by the time he filled out a few worksheets, his anxiety would already begin to diminish a bit. This inspired an immediate sense of hope for him because he saw results happen quickly.

BUILDING RESPONSIBILITY

Creating personal accountability on a daily basis is another important aspect of reducing anxiety. One way you can accomplish this is to commit to take more responsibility for your life by structuring your time and energy wisely. It may sound obvious and trite but after many years of treating patients with anxiety, I've recognized how many of their symptoms are directly related to a lack of structure and integrating meaningful activities into daily living.

In Chapter 6 I discussed how being accountable for your choices and taking responsibility for them can erase guilt and regrets in the future. It's time—right now—to take action and commit to that action. Taking this responsibility prevents you from expecting others to do things for you. It also stops you from relying on external circumstances to somehow miraculously make you feel better. When you're accountable, you learn to feel empowered because you begin to rely on yourself—plain and simple. Plus, as you know, idle time for an anxious mind is fertile ground for extreme negative thinking. In fact it's the ultimate greenhouse effect—the perfect climate for negative thoughts to take root in your head. Your brain wasn't designed to be left unstimulated. You need healthy preoccupation to keep you from falling victim to excessive worry over things you cannot control.

I've often believed that if you stripped human beings of all their responsibilities and daily activities for a few months and had them sit with their idle minds like monks in a monastery, there would be complete pandemonium. People would go stir-crazy, not just out of boredom or from a lack of stimulation. They'd start to unnecessarily overexamine their mortality and freak themselves out about knowing that we'll all die one day, but not knowing when. As a result, people would worry and fixate so much on the inevitable end that their quality of life could diminish drastically. The bottom line is that you have your life to *live*.

And, since you have no control over your mortality, it's important to focus on what you *do* have control over, and that is right now. Your mind needs structure, structure, structure. Did I mention that it needs structure? When you provide that for yourself, your anxiety can be controlled. Obviously, having too much idle time doesn't mean that you'll always be fixated on death, but instead you may overmagnify or obsess over problems that wouldn't be so significant if you were preoccupied with healthy here-and-now things to do. So the next time you complain about having to floss your teeth or take out the trash, remember, even seemingly mundane tasks or chores help you reduce anxiety by not overfixating on everyday issues.

The following exercise can help you preempt the boredom trigger and the mortality freak-out trigger by planning out structure every day so you have very few pockets of unstimulated time. Every evening before you go to bed, take out your Daily Accountability Diary worksheet and begin to map out what you're going to do tomorrow, by the hour. If you have a job or are getting an education, you can still blueprint what you'll work on throughout the day. Then, plan your evenings the same way. If you aren't currently working or in school, then your focus is a little different. Plan out your day by the hour as best you can. If you can't think of things to do, project out what you'd like to do or what you hope you'll get to do. Putting it out there creates some accountability, and you may actually do it if it's in writing.

Consider things like taking a walk, meeting a friend for coffee, going to buy groceries, doing laundry, cleaning your living space or going to see a movie. Use the worksheet that follows to record your plan. Copy it into a notebook or in a document on your computer. Each box under a specific hour is labeled as Activity and Work/Chore. Clearly distinguish between those two and separate them from each other. Fill the boxes from when you wake up to when you go to sleep. Use the samples for Monday in Exercise 4 as a guideline.

DAILY ACCOUNTABILITY DIARY:
EXERCISE #4

MONDAY (a sample)

9:00 a.m.

Activity: Taking a walk with the dogs to the local park for 30 minutes.

Work/Chore: N/A

10:00 a.m.

Activity: N/A

Work/Chore: Cleaning up the garage and making room for storage space.

11:00 a.m.

Activity: Meeting a friend for coffee.

Work/Chore: N/A

Noon

Activity: N/A

Work/Chore: Taking my laptop to do research after meeting my friend.

TUESDAY

9:00 a.m.

Activity:

Work/Chore:

10:00 a.m.

Activity:

Work/Chore:

11:00 a.m.

Activity:

Work/Chore:

Noon

Activity:

Work/Chore:

CONTINUED ON PAGE 204

DAILY ACCOUNTABILITY DIARY

WEDNESDAY

9:00 a.m.   **Activity:**

Work/Chore:

10:00 a.m.   **Activity:**

Work/Chore:

11:00 a.m.   **Activity:**

Work/Chore:

Noon   **Activity:**

Work/Chore:

THURSDAY

9:00 a.m.   **Activity:**

Work/Chore:

10:00 a.m.   **Activity:**

Work/Chore:

11:00 a.m.   **Activity:**

Work/Chore:

Noon   **Activity:**

Work/Chore:

FRIDAY

9:00 a.m.   **Activity:**

Work/Chore:

| **10:00 a.m.** | Activity: |
| | Work/Chore: |

| **11:00 a.m.** | Activity: |
| | Work/Chore: |

| **Noon** | Activity: |
| | Work/Chore: |

SATURDAY

| **9:00 a.m.** | Activity: |
| | Work/Chore: |

| **10:00 a.m.** | Activity: |
| | Work/Chore: |

| **11:00 a.m.** | Activity: |
| | Work/Chore: |

| **Noon** | Activity: |
| | Work/Chore: |

SUNDAY

| **9:00 a.m.** | Activity: |
| | Work/Chore: |

| **10:00 a.m.** | Activity: |
| | Work/Chore: |

| **11:00 a.m.** | Activity: |
| | Work/Chore: |

| **Noon** | Activity: |
| | Work/Chore: |

As you get into the habit of filling in the worksheet for the next day, you may begin to notice that you feel more grounded in day-to-day living, which also helps keep anxiety down. This exercise helped me a lot during my trying days with anxiety. Structure and accountability every day made me focus on something other than my racing thoughts. My recovery became my job and I didn't want to fail. In few words, it gave me my life back by forcing me to work harder at creating my own reality for each day.

Chapter 12

MAKING PEACE
WITH YOUR ANXIETY

Surrendering the Fight,
Not the Cause

*"The real voyage of discovery consists
not in seeing new landscapes,
but in having new eyes."*

—Marcel Proust

Making peace with your anxiety so you can stop fighting it is a concept used in everyday mindfulness training that was first introduced to me years ago. When I first heard it, I thought it was ridiculous. Others might be capable of "making peace" with their anxiety, but not me! I saw myself as terminally unique. My anxiety was just too intense to submit to it. Besides, I firmly believed that if I ever gave up the battle and surrendered to my symptoms, a torrent of even greater and more menacing anxiety attacks would overwhelm me and I'd become incapacitated. Then I'd go crazy and lose my mind. As long as I held tight to my tenuous grasp on what was left of my sanity, I thought I was safer. But eventually I realized that the very same tenuous grasp I held fast to and the exhausting effort I put into it every day made me feel worse.

TAKING THE STING OUT OF ANXIETY

Years later, I discovered that in the process of using mindfulness properly, surrender doesn't mean giving up or allowing myself to be overtaken by anxiety. I learned that if I let go in small increments and allowed myself to be present in the anxiety for brief periods of time, I developed more tolerance to my symptoms as I began to see that I could endure the pain. Getting used to my anxiety meant my brain was getting used to the fear, and it allowed the fear to decrease in strength. Getting more comfortable with my symptoms helped me build emotional tolerance and also raised my distress threshold. And I learned that letting go and surrendering did not mean throwing in the towel. It taught me that I could work on my anxiety and get well by rational means and it's okay to try something different. I didn't have to fight anymore.

The idea of not feeding my anxiety and giving it so much power helped a lot. It allowed me to not believe and give so much importance to every anxious thought I entertained. This led me to understand that, for now, anxiety was a part of me and I needed to let it run its course. In her book, *When Things Fall Apart,* Pema Chodron talks about a young warrior who was taught that she had to do battle with fear, which she didn't want to do. Her teacher insisted she fight. Before battle, the student warrior felt small next to fear, which looked big and angry to her. She asked fear for permission to go into battle. Fear thanked her for showing respect by asking. The warrior asked how she could defeat fear. Fear said that when it uses tactics to get the warrior unnerved, she'll do whatever fear says. But if you don't do what it says, it has no power.

Ironically, the warrior defeated fear by not fighting it. She put her weapons down and instead of engaging in combat chose not to listen to the words fear uttered. This story illustrates the kind of inner dialogue I encourage my patients to begin having about their fearful, anxious thoughts. Responding to your thoughts is better than running from them. As in the previous chapter, this type of internal discourse also helps to transform the anxious experience of being a victim of your thoughts to you being more of an observer of them—and gaining power over them. By successfully resolving a major conflict with peaceful tactics, the warrior averted a lengthy battle, which could have had costly results. By being rational, she chose intellectual methods over emotional ones.

Anxiety management doesn't have to be a struggle. It's sometimes about letting go and surrendering. If you commit to the possibility that looking at things differently might ease your anxiety, the potential results are endless. Therefore, a good goal is to not fight or walk away or quit or give up. Instead, view the same feelings with new eyes. Wayne Dyer says, "When you change the way you look at things, the things you look at change." So, if you change how you look at fear and anxiety, then those emotions can start to change so you can allow yourself to feel better.

SURRENDERING

As a psychotherapist I try to guide patients to find and stay on their path to self-actualization. Since I believe that all people are capable of this goal, as a facilitator in their life journey I help them learn how to remove impediments that get in their way, including anxiety and fear. Surrender is necessary in the process of being a rational warrior because you must be courageous as you sever any negative behavior that's been a barrier to your growth. You need to surrender negative behaviors learned from your past that you still struggle to break free from since this behavior still gives you the illusion of safety and security. This isn't simple. No one likes to surrender anything. But it's necessary if you want to feel more relaxed.

In this context, surrender doesn't mean to give up, concede defeat or become complacent about what's going on. It also doesn't mean grovel, submit or let others have their way with you either. It simply means that you choose to make peace with anxiety by giving up the wretched fight. But at the same time, you don't give up your goal to grow and not let anxiety rule. Just like the young warrior put down her weapons and surrendered old defense mechanisms, even if they sometimes bring temporary relief, you can, too. Surrender means letting go of behavior that's done in an attempt to create safety but instead impairs your ability to change.

Surrender can be effective when you do it by using rational tactics. To assist in the practice of surrendering in rational ways and stepping into the unknown, read aloud the affirmations that follow and let them help you process how to surrender to your advantage. They can also remind you of rational ways to view your anxiety. Use them as often as needed. If any really strike a chord with you, write them down and put them where you see them often.

- Surrender doesn't mean giving in or conceding defeat to my anxiety.
- Surrender doesn't mean groveling or submitting to my anxiety.
- It's okay to feel some anxiety occasionally.
- It's okay to let my guard down and trust that I can handle this.
- Being present with my anxiety from time to time helps build emotional tolerance.
- I don't have to listen to what my anxiety tells me.
- I don't need to control my environment to feel safe.
- Feeling anxious is an opportunity for me to begin a dialogue with it.
- Whenever I'm anxious, I'll stay in the intellectual mind instead of the emotional one.
- I'll allow anxiety to run its course.
- I'll try not to believe everything I think.

DEVELOPING A SPIRITUAL ALIGNMENT

Spirituality is derived from the Latin word *spiritus,* which means to breathe. In itself it's a very personal and subjective practice that has a different meaning for everyone. A very basic way to define spirituality in the context of this book is that it's related to your consciousness, your awareness of yourself and the quality of your relationships with others. Spiritual alignment helps you discover the essence of who you really are as an individual so you can align that core with your personal values.

When you have a spiritual "aha!" moment, or a mind-shift epiphany like the young warrior did in the story, it opens your eyes to the new light that has been shed on a situation. You might even see it as gaining wisdom because, in that very moment, you've learned something pivotal. Your depth of vision will profoundly expand and you'll feel stronger and more confident. Many of my patients have reported that when they have these moments, they feel less alone and more connected to others. Finding your spiritual alignment can support you in managing your anxiety.

You can look at many concepts in this book as being in alignment with the literal definition of spirituality. Being able to breathe can also be an analogy for acquiring the freedom to unburden yourself from the suffocating shackles of anxiety and fear. Balanced thoughts that are reflective and not reactive and involve process orientation lead to behavior that could be considered part of a spiritual practice since it puts you in touch with how you feel and guides you to handle situations in less anxiety-producing ways. Challenging the consensus reality you've always followed and redefining fixed beliefs you've held on to for many years means that you look inside for changing your life. Even realizing that you have the freedom and power to alter your behavior by becoming a more accountable person brings you back to your inner core.

Nurturing your spiritual alignment can make you feel supported and much stronger, which allows you to be more secure about making the changes needed to manage your anxiety. When you're not anxious and you're aligned with the balanced thinking I've discussed throughout this book, you'll feel lighter and able to breathe more easily. You don't need control over anything or anyone, and you alone are responsible for your happiness. Therefore, you are free.

MY AWAKENING

The first time I connected with a deeper awareness of my own anxiety was in my early twenties, after beginning therapy. My anxiety was excessive at the time—between a 7 and 7.5 on the scale. I was unable to work and do the things I wanted to do. My quality of life was poor, to say the least. Although I didn't know it at the time, the main reason for my dysfunction was that I was desperately resisting my symptoms. I did everything in my power to avoid any discomfort. My threshold for pain tolerance was very low. I kept repeating, "I can't stand it; I can't stand it!" I was unaware of the concept of letting go as a means to soothe anxiety, so I naively kept up the fight. I was basically running from myself, but I had no clue that I was doing so.

One day in therapy, my anxiety had spiked to nearly a 10 after a scary argument with my father the night before. In those days I had no voice against my father's wrath and felt oppressed and powerless to go up against him. My suppressed rage turned into anxiety. Unsteady with angst, I began pacing in my therapist's office as I often did when my symptoms escalated. I was restless, my mind was spinning and I began clawing at the cuticles on my fingers. Sometimes if the panic was really intense, I even walked (or ran) out of the office for a few minutes to get air. Or sometimes I didn't return to the session and would call my therapist from the street and apologize.

That same day, as I paced back and forth, with my symptoms mounting every second, my therapist changed her intervention strategy. After she validated my distress, she suggested, "I'd like you to try and be present with it today." She had never asked me to do this before because she knew I hadn't been ready for it. So she waited until the right moment and knew it had to be right then. After I called her insane for even considering such an absurd idea, she kindly and respectfully ignored my refusal and invited me again to try. Suddenly, it dawned on me that up until this point, I had avoided being present with my anxiety. After a whole life of not facing it, I knew I could so easily escape it yet again.

My instincts told me to ignore her and flee as I had always done. But for some reason the phrase "be present with it" hung in my mind like a dare. I was frozen yet mysteriously intrigued by the unknown. I vacillated since I still could not bear to just be with the intense anxiety I was experiencing. I teetered between running out the door or staying inside and going absolutely crazy or psychotic. Reluctantly, I returned to the couch and sat down. My inner world was spiraling with all kinds of catastrophic thoughts about how I was going to die right there in her office. The more I spun, the worse the symptoms got. I thought I was going to explode. Fear and panic overwhelmed me as I tried to take control using my usual methods.

Then I heard her voice again. She said, "Right now, your desire for this moment to be different is causing you more distress than you think. Let it wash over you. Accept that you have anxiety at this moment, just this moment." She paused for a beat, and then added, "And I want you to close your eyes and focus on your breathing for as long as you can…nothing else, just your breathing. Can you do that?" Even though I was churning inside, I trusted her enough to step off the proverbial ledge. For a brief period, I let it wash over me as she requested. I dropped my weapons and surrendered.

Tiny increments of surrender were all I could muster at the time, but it was all I needed to break the ice. I performed the breathing exercise much like the one I share in the following section and realized that focusing on the inhaling and exhaling of my breathing as she prompted me to do was crucial for the redirecting of my thoughts. Every deep breath I took expanded my consciousness enough to open up space around the negative reflex of escaping and avoiding. It wasn't seamless or easy, but I did it. After executing the breathing commands for about ten minutes, I noticed that my anxiety dropped to about a 5. I felt slightly encouraged feeling the difference, even though I was battered and drained by my symptoms.

The escape artist inside me never wanted to be present with my anxiety ever again. But I knew I had to persist in doing it. For the next few months, I continued that intervention ritual in my therapist's office whether I was anxious or not. I even began doing it at home. The awareness that my resistance to the anxiety was causing my symptoms to exacerbate was monumental. My therapist helped me to slow down my racing thoughts and alter the hard wiring I'd always had of running from anxiety. I accepted the process of rewiring how I managed it. Like the young warrior in the story, in a complicated and puzzling way I coped with my surging anxiety by doing the opposite of what my instincts said to do and made peace with it. It was a determining moment in my life that I'll never forget. In a sense it was an awakening of myself.

DAILY MINDFULNESS EXERCISE

I'm going to share a daily mindfulness exercise that will expand your ability to align your consciousness with the spiritual outlook discussed earlier. Hopefully it will help you learn to breathe easier when you need to. It's another step in the process of making peace with your anxiety. This exercise will also use the affirmations and replacement thoughts learned earlier in this book. Mindfulness exercises open up the deeper areas of your consciousness that are sometimes unavailable to you because you're stuck in what Deepak Chopra calls "the known"—your dependence on the "prison of past conditioning." In this instance, the past conditioning can be clinging to fixed beliefs systems, thinking in a dualistic mind and going along with a consensus reality.

Don't save this exercise just for moments when you feel very anxious or stressed. It's intended to help you build your mind-conditioning skills over time and needs to be done regularly. It will guide you to build up the self-regulation or self-relaxation muscle in your mind that needs a lot of training. Consider it like going to the gym for a workout, except that this type of workout can be done in the privacy of your own home or anywhere for that matter. By performing this exercise regularly, you'll also begin to lower your everyday anxiety baseline, which is probably very high. If you suffer from anxiety, your baseline may be hovering between a 7 and a 9. Lowering it to a manageable 4 or 5 on the scale would be the most realistic.

I suggest that you try to do the exercise once a day, preferably in the morning, and then see how your day goes. In time, I bet you'll see a difference—feeling more relaxed and less burdened by the stressful factors that impede your life. This exercise shouldn't take more than fifteen to twenty minutes to complete and it is well worth your time.

Find yourself a comfortable place to sit in your home or at work, in a quiet room or space where you won't be disturbed for a while. Allow your back to be upright either against a wall or the back of a chair. Let your arms and hands rest on your lap or your knees or wherever feels comfortable to you. If you sit in a chair or couch, allow your feet to rest flat on the floor. If you sit on the floor, it's best to fold your legs beneath you. Before you begin, rate your current anxiety on a scale of 1 to 10 like before. "Currently my anxiety is a ___." Then close your eyes and begin.

Breathing

The first step in doing the meditation is to tune into your breathing, which is an important sensation of the body that you may never pay attention to. Place your right hand over your stomach. While holding it, take a slow and deep abdominal breath that lifts both your stomach and chest. Let your breath push your stomach out until you feel it expand. Feel both your chest and stomach rise with your hand still on it. Hold the breath for five seconds, and then exhale slowly through your mouth. Repeat this five times.

Whenever you do this abdominal breathing, keep your eyes closed and focus on the image of your stomach and chest rising rhythmically. Pretend you're outside of yourself looking in as an objective observer. You could also focus on how the air feels passing through your nose and down into your lungs or imagine you are the air itself as you enter your body and travel through your nose and then exit through your mouth. The intention is to focus intently on any part of your breathing that feels comfortable in order to help you fine-tune awareness of your body. The more you repeat this breathing every day, the better you'll become at refocusing your negative thoughts in the future.

It may seem dull to concentrate on doing something simple like breathing, but mindful breathing slows your mind down and relaxes it. Ultimately it will help ground you in the here and now, which is accessed merely by the conscious acknowledgment of your breathing. Try to imagine that with each exhale, you're letting go of all your negative thoughts while breathing in soothing, positive thoughts. It's important to remember that when your mind wanders and you stray from your breathing, nothing is wrong. It's normal to let your thoughts go to something else. Actually, that's what is supposed to happen. When it does, simply bring your attention back to your breathing and refocus. Every time you bring your mind back and refocus, you build that mindfulness muscle, and it gets stronger and stronger.

Try to picture your thoughts as a helium balloon that's tied to a string in your hand. Every time you loosen your grip on the string and the balloon starts to float away, you can pull it right back in. It works the same way with your breathing. The balloon starts to drift and you pull it back in; it drifts again and you pull it back in. Every time your thoughts drift away from your breathing, you can pull them back in. This is what you'll do for this exercise—just you and a helium balloon.

Affirmation

After each round of five abdominal breaths, add an affirmation. When you do your last exhale of a round, quietly whisper the following surrender declaration to yourself:

In this moment, I let go of the fight. I can handle anything that comes my way.

It's okay to feel some anxiety every now and then.

Then return to another set of five abdominal breaths, still concentrating on the sensation of your breathing, and whisper to yourself:

> *In this moment, I don't have to believe what my anxiety tells me.*
>
> *If I get anxious, I will access my rational mind instead of the emotional mind.*

Then return to another set of five abdominal breaths, still concentrating on the sensation of your breathing, and whisper to yourself:

> *In this moment, I don't need to control my environment to feel safe.*
>
> *Surrender does not mean putting myself in danger. Everything is okay.*

After you finish the affirmations, the exercise is completed for the morning. Remember, say one affirmation after every five breaths. Try to do three sets of five breaths and affirmations each morning. When the exercise is done, rate your current anxiety on the scale of 1 to 10. How low is it? Over time you'll see that your anxiety rate may decrease after every round of exercises. It's important to keep a record of the scale ratings so you can observe your progress. Be patient—it may take a few weeks to start seeing the numbers go down. Remember, use a process orientation, not one that seeks immediate results, and allow it time to work.

THE IMPORTANCE OF AWARENESS

I would like to share one final illustration of how you can be mindful by surrendering the fight but not the cause. In his eminent book, *Memories, Dreams and Reflections*, Carl Jung described a dream that inspired me to pay close attention to the importance of having mindfulness, despite insurmountable odds and obstacles. He said that his dream both frightened and encouraged him. It took place during the night and he was walking alone. His progress was slow and painful because of a strong wind and dense fog. His hands held a tiny light tightly, which he worried would blow out at any time. Jung felt that everything hinged on not letting the little light go out. Suddenly it felt like something was coming up behind him.

In the dream Jung turned and saw a large, dark figure behind him. Despite the terror he felt, he knew he must keep his light going through the night, despite other dangers that might lurk. Then he woke up and recognized that the dark figure was his own shadow, distorted by the fog, and created by the little light he carried. He also recognized that that light was his consciousness, the only light anyone has. In this dream, the insight into the depth of his own awareness helped Jung to see that if there's light, there must be dark. One can't exist without the other. Light has no meaning if you can't understand or perceive its opposite—darkness. And vice versa.

This dream is also a reference to the perils of letting a dualistic mind control your thinking. There are no extremes in life, except when you create them in your mind. Even the power of light is always tempered by some dark. However, the tangible beauty in the dream and how it relates to the many concepts I've laid out is that the shadow Jung describes that's following him comes from his own body blocking his little light. The shadow he fears is fictitious. It's not a circumstance he has no control over. There's no one to fear. Whatever he creates, he can recreate, or change. The lesson?

I can create my own reality.

Therefore, the message can be interpreted that perhaps your own negative thinking and self-defeating beliefs obscure your consciousness, which is your light, and cause you to be frightened. You can learn to live with it and accept the shadow, or you can run from it. In this case, Jung realizes that despite his terror, he knows he must keep the light alive and preserve it no matter what. He even suggests that the light is small and fragile compared to his shadow, but nonetheless, he must remain mindful and become the true keeper of his flame. Perhaps then he becomes a rational warrior because he's encouraged by the fact that since *he* is generating his own fears, then *he* has the ability to let them go, just as *you* generate your own fears, and *you* have the power to let them go.

Jung's story is also an example of the spiritual alignment discussed earlier. The dream helped him surrender to the dark because he knew it couldn't harm him. This pivotal shift in his awareness elevated his consciousness to a greater understanding of himself. Maybe it even helped him breathe easier after he made that connection. Such epiphanies or spiritual moments are very encouraging to experience, especially when you've struggled with many years of fixed thinking about your anxiety. These are no miracles. Instead, they're basic consciousness shifts that you can generate for yourself if you commit to staying open to them.

My hope is that you'll consider your "light" as a central part of your recovery from anxiety. When all else around you feels chaotic, you can use your mindfulness skills to come back to the light—your only light. The power to manage your anxiety is in *your* hands, and how much you can become aware of your thoughts and how they create anxiety. I came from a place where my anxiety sometimes was at 10 points on the scale. I did everything I could to avoid facing it, which meant I didn't want to take responsibility for what made me anxious. So I suffered a lot for many years. If what I've shared worked for me, it can work for you.

I hope that you'll use my tools to take charge of the way you perceive your life, the people in it and the situations that come up. When you do, you'll enjoy having less anxiety. I can suggest things to do but the power is only yours. *You* can change your situation or let it continue to stress you out. By using the tools in this book, you can retrain your anxious mind to relax and perceive life in a less anxious light. I wish you the great joy that managing anxiety can bring!

Afterword

LIVING IN
THE PANDEMIC ERA

The New Age of Anxiety

When *Retrain Your Anxious Brain* was first published in 2014, no one had any idea how drastically the world would change in the ensuing years, and especially in recent months. The coronavirus pandemic not only upended our lives and kicked us out of our comfort zones, but it put us on a fast track to a new age of virtual existence, depersonalization, and persistent anxiety.

So much has changed. The COVID-19 safety protocols and restrictions of wearing masks, quarantining, and social distancing profoundly altered our lifestyles. It changed how we access health care, how we work, how we go to school and learn, and how we interact with fellow humans.

Even now, way beyond the one-year mark of the beginning of the 2020 pandemic, American lives are still profoundly impacted in so many ways.

And, though millions of people are vaccinated and COVID-19 precautions are significantly eased or removed, with businesses and schools reopening, we still remain vulnerable to new variants and strains of the virus that could trigger another rise in infection rates in the future. Booster shots may help, but many scientists say we will never be completely out of the woods. The pandemic era and the new age of anxiety are here to stay.

With the prevalence and persistence of the pandemic, we are especially witnessing an alarming rise in cases of serious mental illness like major depression, severe anxiety, substance abuse, and various forms of trauma. The reality is we may not see the full effect of the pandemic on mental illness until Americans have had time to assess their losses—losses such as jobs, businesses, homes, life savings, and the deaths of loved ones and friends. We will also have to reckon with other losses that are not easily quantifiable, like the diminution of our basic freedoms, the forfeiture of certain traditions, the curtailing of celebrations, and vanished time that can't be recovered.

COMPLICATED GRIEF

High anxiety and the buildup of "complicated grief" will hit America like a tsunami in the near future. Complicated grief causes people to get "stuck" in mourning and exist in an extended state of bereavement. It's like being frozen in a constant new norm of angst, numbness, and rumination. According to the Mayo Clinic, "For some people, feelings of loss are debilitating and don't improve even after time passes. In 'complicated grief,' sometimes called *persistent complex bereavement disorder*, painful emotions are so long lasting and severe that you have trouble recovering from the loss and resuming your own life."

The pandemic is more than a tragedy; for many, it is a catastrophe. In a tragedy, some lessons can gradually be gleaned from the event, which helps to alleviate the grief and, over time, stabilize the pain we feel. But with a crisis like COVID-19 that becomes a catastrophe, we benefit little and understand less.

When the nightmare fully ends and we wake up, many of us will exhale a collective sigh of relief and go on with life as usual. But for those of us who find ourselves standing amid emotional wreckage and facing irretrievable casualties, it will feel more like a catastrophe, aggravating our anxiety and possibly creating a mental health crisis.

In my psychotherapy practice, I have seen some of my patients confronting catastrophe. Some reported that what they feared most was that the constant exposure to toxic worry and distress would take them down into an abyss of psychiatric incapacitation. They dreaded "losing it" and experiencing what we used to call a "nervous breakdown." In other words, they worried about *worrying*. Worry and fear live off of the tyranny of uncertainty.

EMOTIONAL TRAUMA

The pandemic caused massive emotional trauma, an often overlooked condition. Exposure to adverse events can shatter our sense of safety and security. Since March 2020, the continuous levels of emotional trauma and secondary trauma that individuals have been exposed to is immeasurable. Trauma symptoms do not manifest only after a single adverse event occurs. They can also develop over time, where the overexposed human psyche is stretched to its limit, unable to process the troubling circumstances.

Subsequently—because trauma is deeply encoded in the senses— it can leave one with a dangerously changed, over activated central nervous system, causing debilitating symptoms such as anxiety, severe depression, numbness, confusion, nightmares, suicidality, and more.

In his definitive book on trauma, *The Body Keeps the Score*, Dr. Bessel Van Der Kolk tells us, "After trauma, the world is experienced with a different nervous system that has an altered perception of risk and safety." This means that the body and mind remain fixed in constant "threat response" or survival mode, which is an exceedingly distressing state to exist in for prolonged periods of time. The mind, already overwhelmed, cannot always handle it. Secondary trauma, also known as vicarious trauma—when someone witnesses adverse events happening to other people—has been prevalent since the onset of the pandemic. Witnessing traumatic events can have as much of an impact as personally experiencing them. Both direct emotional trauma and secondary trauma can cause similar incapacitating symptoms.

"If an organism is stuck in survival mode, its energies are focused on fighting off unseen enemies which leaves no room for nurture, care and love," says Van Der Kolk. Severe anxiety can manifest and stimulate a potentially fatalistic outlook on life.

The pandemic has been one long terrifying adverse event that has triggered individuals to suffer from a new kind of trauma we have never seen before. Its effect has been compounded by other anxiety-inducing phenomena in our lives. In 2020, many cities also experienced violence, protest, and social unrest. Many individuals felt threatened by the power of the state; people across the political spectrum saw the government, and society itself, as suspect. In California severe wildfires destroyed hundreds of homes and displaced thousands of people. Then, in the fall of 2020, the stress of election season rolled in and turmoil and divisiveness among Americans resurfaced with elevated anxiety levels.

Let's also not forget about the many individuals with previously existing mental health conditions for whom long-term exposure to new trauma like the pandemic can exacerbate symptoms. For example, people with depression may experience a kind of "double-depression" due the trauma of the pandemic piled on top of their preexisting condition. And individuals who previously suffered from anxiety disorders witnessed their fears and catastrophic thinking skyrocket.

So, the combination of worrying about an uncertain future, fearing death, perhaps suffering unemployment and/or financial hardship, and enduring isolation and loneliness due to social distancing directives all add to the burden on any person's mental constitution. Plus, social distancing, although necessary to prevent the spread of the virus, deprives us of vital human connection.

PSYCHOLOGICAL EFFECTS OF SOCIAL DISTANCING

Human beings are social creatures. We have evolved over the millennia into unavoidable interdependence. That's how we cope with adversity, tragedy, and calamitous events. Social distancing challenges our fundamental human instinct to jointly seek support in large groups—whether family, friends or neighbors. We don't live in a vacuum. We need each other.

In the past, when humans have faced disasters, we have relied on a sense of mutual experience and community to process the events. The 9/11 attacks in 2001 are an excellent example. Despite the horrific events of that day, Americans swiftly came together in the aftermath and remarkably coalesced in the shared grief. Collectively, the human spirit is more apt to be inspired when we rally together for a common cause, especially if it's a life-and-death one.

Therefore, we have to ask an important question: Did our current forms of virtual communication make up for the drawbacks of social distancing? Did texting, phone calls, email, and video chat platforms like Skype, FaceTime, and Zoom truly help the situation? And if so, was it enough?

We are very lucky to live in an era where technology allows us to see our friends and family from a distance. However, these modes of communication cannot replace face-to-face interactions. They cannot replace the loving energy that is transmitted *live* from one person to another. They cannot replace warm facial expressions and especially tender, compassionate touch. Yes, we are a high-tech society but we are also a high-touch species.

Long-term exposure to social isolation may cause such deep emotional numbing and disconnection that we can lose both a sense of reality and identity.

In her book *A Guide to Crisis Intervention*, Dr. Kristi Kanel, PhD, says, "Numbing may also arise as individuals interact with others only via Zoom and social media rather than in person. Many people talk about Zoom fatigue, which may be similar to combat fatigue. When people feel fatigued and powerless, they often give up and exist just to survive rather than work to escape the situation." In other words, individuals may not even be aware of the significance of the disconnection that comes from virtual communication, and consequently may not take their mental health needs seriously.

We don't yet know the effects of short-term social isolation. It has not been studied fully. However, we do know that social isolation over a long period can lead to health conditions like heart disease, high blood pressure, Alzheimer's, dementia, and even death. Chronic loneliness can also impair our immune system and leave us vulnerable to disease.

I believe that unfolding on top of the COVID-19 crisis, there may be a secondary social crisis that could persist long after the pandemic retreats. Unfortunately, I don't believe that we as a society are fully prepared for it yet.

We still dehumanize people suffering from mental illness and regard them as inferior. We have indeed made great strides in the last few decades, but due to continued detrimental attitudes and misconceptions about mental illness—inspired by negative portrayals in the media, stereotyping, and fear—we still have a long way to go. Currently, we may not have the resources in place to help the multitudes of Americans who might spiral out of control before our eyes.

In early 2021, Kim Tingley of the *New York Times* reported that "there is evidence suggesting that significantly more people have thought about ending their lives during the pandemic than in recent years. In August 2020, the Centers for Disease Control released the results of a nationwide survey conducted during the last week in June: More than 40% of those who responded reported symptoms of anxiety, depression and increased substance use, in addition to other struggles. And more than 10% said that they had seriously considered suicide in the past 30 days."

Ten percent is roughly 32 million people, and that just includes the individuals that responded to the survey. Since that survey, the numbers for mental illness cases due to the pandemic have risen. A survey by the Centers for Disease Control from December 2020 to April 2021 found that 42% of Americans were suffering symptoms of anxiety or depression. And only 11% reported experiencing these symptoms before the pandemic began.

But despite the prevalence of these mental health issues, many Americans are often still reluctant to address their depressive or anxious symptoms.

Interestingly, some individuals who have long histories of anxiety and excessive worry are managing quite well in the pandemic era. In their minds, a catastrophic event like the spread of the coronavirus levels the human playing field. They feel less stigmatized by their affliction—because now everyone is anxious, too. They also feel like families and friends may finally understand their suffering better, which temporarily eases the shame they are used to feeling. "Now we're all in the same boat," one patient told me.

In addition, there are others who are for the most part homebodies and perhaps live introverted lives. These groups of individuals are comfortable with quarantining and social distancing. They revel at working from home and not dealing with the awkwardness of socializing with others. The social distancing mandate gives them permission to avoid human contact. This form of social anxiety disorder is more common than some may think.

However, these cases are not the norm. If you are living through the pandemic and the aftermath, you cannot deny the sword of Damocles that swings over our heads. Will the sword ever stop hovering over our frightened lives? It's hard to say, but in a way it's always been there. It's part of the human condition to fear the worst. But since 2020, we have been forced to look up at the sword more frequently.

In the coming months and perhaps even years, the numbers of people needing mental health treatment will grow. We can hope that untreated mental illness will not precipitate other catastrophes. Maybe, with numerous cutting-edge therapeutic modalities that may assist in treating pandemic anxiety, emotional trauma, and depression, we can untangle the complicated grief, save some lives, and return people back to their previous level of functioning.

In the distant future, whether we reflect back on the pandemic as a tragedy or a catastrophe, I am optimistic that this awful experience will make us a stronger and smarter nation and prepare us for any future crisis.

I am excited about the reissue of *Retrain Your Anxious Brain*. I believe it not only helps us deal with anxiety that already arises from everyday life, but will now be an excellent resource for anyone experiencing chronic anxiety due to the effects of the pandemic.

The time-tested, evidence-based therapeutic tools and stress-reducing skills in this book have worked for so many. They can work for you, too.

JOHN TSILIMPARIS, MFT

BIBLIOGRAPHY

Beck, Aaron T. 1976. *Cognitive Therapy: And the Emotional Disorders.* New York: Meridian.

Beck, Aaron T. and Clark, David A. 2010. *Cognitive Therapy of Anxiety Disorders.* New York: Guilford.

Burns, D. David. 1980. *Feeling Good: The New Mood Therapy.* New York: New American Library.

Chodron, Pema. 2000. *When Things Fall Apart: Heart Advice for Difficult Times.* Boston: Shambala Library.

Chodron, Pema. 2001. *The Places That Scare You.* Boston: Shambala Classics.

Chopra, Deepak. 1993. *The Seven Spiritual Laws of Success.* San Rafael, CA: New World Library.

Corey, Gerald. 2008. *Theory and Practice of Counseling and Psychotherapy.* New York: Brooks Cole.

Ellis, Albert and Harper, Robert A. 1961. *A Guide to Rational Living.* North Hollywood, CA: Wilshire Books.

Hanh, Thich Nhat. 2001. *Anger: Wisdom for Cooling the Flames.* New York: Riverhead Books.

Hanh, Thich Nhat. 2002. *No Death, No Fear: Comforting Wisdom for Life.* New York: Riverhead Books.

Jung, Carl G. 1963. *Memories, Dreams and Reflections.* New York: Pantheon Books.

Kanel, Kristi. 2018. *A Guide to Crisis Intervention* (6th ed.) Pacific Grove, CA: Brooks/Cole.

Mayo Clinic. *Complicated Grief.* https://www.mayoclinic.org/diseases-conditions/complicated-grief/symptoms-causes/syc-20360374.

Sartre, Jean-Paul. 2007. *Existentialism Is a Humanism.* New Haven & London: Yale University Press.

Tingley, Kim. "Will the Pandemic Result in More Suicides?" *New York Times Magazine*, January 21, 2021. https://www.nytimes.com/2021/01/21/magazine/will-the-pandemic-result-in-more-suicides.html.

Tolle, Eckhart. 2003. *Stillness Speaks.* California: New World Library.

Van Der Kolk, Bessel. 2014. *The Body Keeps the Score.* New York, NY: Penguin Books.

Yalom, Irvin D. 2005. *The Theory and Practice of Group Psychotherapy.* New York: Basic Books.

Yalom, Irvin D. 2008. *Staring at the Sun.* San Francisco: Jossey-Bass.

Zettle, Robert D. 2007. *Act for Depression.* Oakland, CA: New Harbinger Publications, Inc.

INDEX

M

N

O

P

ACKNOWLEDGMENTS

Many thanks to my agent, Linda Konner, and my coauthor, Daylle Deanna Schwartz, for their commitment to the book, and to Sarah Pelz and Rebecca Hunt at Harlequin for sharing my vision with this project.

A grateful acknowledgment to my early mentors as well, Claire Ciliotta, PhD, Eve Siegel, MFT, and Marsha Jacobs, MFT. This book could not have been written without them.

Also a big thanks to all the patients I have had the privilege of learning from and helping. This book is dedicated to all of you.

JOHN TSILIMPARIS, MFT